Understanding the Construction Client

David Boyd
Professor of Construction
University of Central England

and

Ezekiel Chinyio
Senior Lecturer
University of Central England

Blackwell
Publishing

Editorial offices:
Blackwell Publishing Ltd, 9600 Garsington Road, Oxford OX4 2DQ, UK
    Tel: +44 (0)1865 776868
Blackwell Publishing Inc., 350 Main Street, Malden, MA 02148-5020, USA
    Tel: +1 781 388 8250
Blackwell Publishing Asia Pty Ltd, 550 Swanston Street, Carlton, Victoria 3053, Australia
    Tel: +61 (0)3 8359 1011

First published 2006 by Blackwell Publishing Ltd

ISBN-10: 1-4051-2978-6
ISBN-13: 978-1-4051-2978-7

Library of Congress Cataloging-in-Publication Data

Boyd, David.
    Understanding the construction client / David Boyd and Ezekiel Chinyio.
        p.   cm.
    Includes bibliographical references and index.
    ISBN-13: 978-1-4051-2978-7 (pbk. : alk. paper)
    ISBN-10: 1-4051-2978-6 (pbk. : alk. paper)   1. Communication in the building trades.
2. Construction industry—Customer services.   3. Consumer behavior.   I. Chinyio, E.
(Ezekiel)   II. Title.
    TH215.B69 2006
    690.068′8—dc22
2006005837

A catalogue record for this title is available from the British Library

Set in 10/13 Palatino
by Graphicraft Limited, Hong Kong
Printed and bound in India
by Replika Press Pvt Ltd, Kundli

For further information on Blackwell Publishing, visit our website:
www.blackwellpublishing.com/construction

Contents

Foreword

by the Chair of the Construction Clients' Group

This book is what I, and a great many in the construction industry, have been waiting for. The behaviour of clients is one of the most – if not the most – important ingredients in improving the performance of the construction industry. Therefore, understanding clients, by understanding what underpins and prompts their attitudes and actions, becomes critically important to the supply chain in collectively taking the construction industry forward.

At the Construction Clients' Group (CCG), which represents a wide range of customers of the construction industry, we need to understand the industry as a whole, not just the client side, an understanding that we use to encourage clients to procure construction in support of improved performance across the construction industry. Performance improvement must include better value for clients, sustainable outcomes, and a prosperous and successful construction industry. The CCG's actions are a continuation of the client leadership theme, which our two predecessor bodies, the Construction Clients' Forum (CCF) and the Construction Clients' Charter (CCC), both pursued.

But this represents only one facet of the CCG's role. We must also foster a better understanding of clients among the construction industry. I was delighted, therefore, to be able to support David Boyd and Ezekiel Chinyio in the development of this book. They say it is the construction industry's duty to learn to work with clients better, and that buildings are not about building – not for clients, anyway. In other words, clients see and experience buildings differently from the rest of the industry. Most clients are preoccupied with their specific world of business and the pressures on this. When a client considers a building project, it considers the project in terms of its business, not in terms of the detail of construction implementation. The industry, on the other hand, is concerned with the details of design and construction in relation to its effort or liability and may not see the client's business and business environment as relevant to the industry.

Another issue that is uncovered in this book is that, for most clients, building is a consequence of organisational change, and organisational change creates uncertainty and emotion. Much of this book therefore addresses human psychology. We would make no apologies for that. People are important in organisations, and in most cases one could argue that people embody the organisation. Expectations and perceptions are also covered. As the authors say, if the client's expectation of parts of the industry is that the industry is profligate in spending someone else's money, then the client will be sensitive to any suggestions of additional expenditure, even if it would receive some benefit.

David Boyd and Ezekiel Chinyio recognise that clients differ in their internal organisation, objectives, drivers and values, particularly between public- and private-sector clients. I can vouch for that, having been Chief Executive of the London Borough of Barking and Dagenham from 2000 until 2004, before transferring to the private sector, where I run Leisure Connection, a national leisure operator. I can also verify the authors' claim that the differences are less than they once were. Of course, clients are not homogeneous, and that is recognised in this book.

The true importance of the construction industry to the national economy is appreciated fully by few. As emerging texts such as *Be Valuable* demonstrate, the performance of the wider economy largely tracks the performance of buildings. It is that simple, and understanding clients and breaking down the barriers between the customer and the industry is key to unlocking the potential of construction and with it the wider economy. As explained by the authors, in order to achieve understanding one must have collaboration and joint working. Everyone in the construction industry, and those buying into it as clients, need to foster increasingly close working between clients and the whole supply chain.

In my role at the CCG, I am grateful for the support of all my colleagues. I would like to give special thanks to the CCG's Vice Chair Peter Woolliscroft, who facilitated a number of the interviews underpinning *Understanding the Construction Client*.

I hope that this book inspires you to take a different perspective on the respective roles of the client and construction industry. If it opens up a new angle of dialogue and understanding, then it will have served its purpose.

Graham Farrant
Chair of Construction Clients'
Group, the only body dedicated
to all clients of the construction
industry

Preface

The statement that encapsulates this book is:

Buildings are not about building!

This simple, yet provocative, statement has implications for both clients and the construction industry. If you have not dismissed the book already because of this idiocy then you will be asking yourself, well if buildings are not about building, what are they about? Of course, there is no simple answer and you will have to read the book to find out. But in the meantime let us look at the statement.

The noun 'buildings' certainly represents the end point of the verb 'building'. That is, building as a verb is the means and building as a noun is the end. In this, the action of building delivers a building. But note we are saying buildings are not about building. A building has physical reality and more than that has substance. But that is not enough to define what a building is; what it also has is utility and purpose, indeed it may even have presence. All of these, however, start involving some interpretation by individuals or groups. And so it is this distinction between the physical substance of a building and its meaning for people that starts us on our journey to understand why buildings and building are not trivial, and produce considerable problems. It is also the issue that is most often forgotten. Somehow we feel that the way we see a building is the same as others do. The fact that the physical substance is so concrete, and we can agree what is there physically, seems to lead us to believe we are all seeing the same thing. Buildings are complex and like a kaleidoscope may change at every turn and like a modernist painting can be interpreted differently. They are in the eye of the beholder, indeed they are in the mind and heart of the beholder. This can create a disjunction between the physical reality and the expectation. Buildings can be ascribed with more than their reality; they have a deeper meaning than their substance.

What then of building as a verb. Building is a dynamic activity that is fundamentally about change. In comparison to lots of other business or service actions, which are about maintenance of an activity, building involves a continuous change of activity: conception, design, construction, commissioning and use. It is only at the point of use that we get some stability in activity. Thus building involves a progressive change, one step rests on the previous and is therefore fundamentally dependent on it, i.e. steps are interdependent. In order to engage in this really complex activity, we require some rational planning and management. It is in this that the end (the building, if our assumption is that buildings are about building) requires to be conceived or indeed preconceived. This involves such activities as briefing and design and these again are taken as rational production actions. There is much debate about the degree of rationality in building and different people see this differently, either through personal disposition or through a learnt role. Thus, although the reality is in the end rational, expectations certainly are not. Buildings are conceived and then produced in the future allowing for great disjunctions as reality comes to meet expectation. In all this there is a tremendous emotion. Our expectations are high, but like a gambler we place all on the hope that the building will meet our expectation. Our investment into it is large, which involves a great emotion. If it is deviant there is even more emotion.

The opening statement could less provocatively be 'buildings are not *just* about building'. However, to start to understand clients we must avoid falling into discussing the construction activity and consider more 'what a building means for the client', not its means of production or its physical realisation. Clearly the production of buildings involves change, but more importantly everyone associated with them is changing, in particular the client. The answer then to the question of what buildings are about is:

Buildings are about changing and developing the client!

Thus, the subject of this book and what we say is required to understand clients: that is, how clients develop through buildings, how they perceive buildings and how they perceive and experience the process of building. Most texts on clients spend their time on improving the construction process. They see improved briefing or improved design or improved procurement or improved construction as the means for delivering better buildings for clients. We do not deny that these are important, but they do displace our gaze from what we think is a critical action of understanding the meaning of building for the client. What we are saying in this book is not new, we are just emphasising something already understood in order to develop thinking about building. This emphasis helps us to escape from our (industry and clients') current conceptions about building, which are part of the problem and which cause us to reproduce the difficulties of working with clients. The structure and processes of building embody these conceptions and so embody the difficulties.

What, then, can we do about it? First, and as the title of the book alludes to, we must understand the client. However, this is clearly not sufficient, we must also act differently and the line we are taking is that we should work as a 'process consultant' to the client. This idea relates to management consultancy, but we believe it can help us to work with the complexities of the change taking place in the client. We do recognise the difficulties in this. We are still expected to be technical experts and build buildings; thus, when can we develop the skills? When have we the time to do these extra tasks and when do we get paid for this extra work? Partly, of course, we are saying it is not extra work, it is THE work and our conventional task is made easier and will deliver more satisfactory outcomes for all if we do this. As with all proposed changes in the industry, there is a lot in our history, current structures and the current business environment to prevent it happening. It is this we need to work on.

It may look at times as if we are criticising the client. This is far from our intention. Clients need their problems managed by the industry when they build. Thus, we need to identify and analyse what some of these problems are. However, as we are focusing on client problems, it means that we exaggerate their significance. This does not mean that these problems are the fault of the client; they are, in fact, consequences of the client's world. This does not mean that these problems occur all the time and that clients cannot deal with them. We are merely identifying that there is the potential for such problems to occur that require support from the industry. As problems are context-dependent, it is unlikely that they will occur in the same way. Thus, these problems that we focus on are only indicative examples and it is up to the industry, along with the client, to identify similar problems as they occur in practice.

The ideas for this book came from many years of working with MSc Construction Project Management students at the University of Central England in Birmingham. This course asked students to explore the realities of the industry rather than work on idealised conceptions. The work of Alan Wild with this course and on the ideas surrounding this book is gratefully acknowledged. The students were from clients as well as consultants and contractors. They revealed in their discussions together the inadequacy of the current arrangements that they had to manage and the potential for working differently. This book is a testament to their experience, their reflections and their abilities. Several undertook dissertations about clients: Adrian Wheeler; David Cant; Byron Pountney; Tony Catchpowle; Eamonn Kerr; Steve Gilbert; Kevin Small; Paul Baker; Martin Jarrett; Claire Charman and Linda Cresswell; who provided priceless insights into practice. We thank them all.

The book is based on developing a number of positions on the industry; we would like to acknowledge these sources as fundamental to our inquiry. These are Graham Winch of UMIST on Construction as an Information Processing System; Peter Barrett of the University of Salford on a wider systems perspective of construction and the importance of Facilities Management; and Stuart Green of the University of Reading for his critical sociology of the realities of the industry. In addition, we would like to acknowledge the work of the late Steven Groak, who saw problems of construction as part of the constitution of construction rather than a mistake. There is also a psychodynamic thread which comes from the work of Jean Neumann of the Tavistock Institute, London and the cooperative inquiry work of John Heron. Finally, the complexity work of Ralph Stacey at the University of Hertfordshire allowed us to acknowledge the unknown in the development of the future.

We must also thank the interviewees from the sectors who had to suffer our inquisition. A list of these is presented in the Appendix. We investigated other sectors than those published here but we discovered that it was extremely difficult to encapsulate a sector and this took longer to describe. We limited it to six sectors. The ideas for the final outcome developed through these interviews rather than being preconceived by us. In addition, several read and commented on the sector chapters to ensure their accuracy. The remaining errors are entirely our fault through our misperceptions about the complex world of clients. The industry is extremely keen to work better for clients and great interest has been shown in the book. Also, clients are interested in getting the industry to work better with them and are themselves working on the problem. We must thank the Construction Clients Group for giving us access to this activity, especially Christopher Morley, the Chief Executive, and Graham Farrant, the Chair.

We must also thank the people in our lives who suffered as a result of the book, our work colleagues, our students, but most particularly Gill and Rachel, whose support was critical for its completion.

In the end, we all need to change to survive!

David Boyd
Ezekiel Chinyio
Birmingham
May 2006

1 Clients in Perspective

Today, more then ever before, business is about fulfilling client satisfaction.

Pries *et al.* (2004)

1.1 INTRODUCTION

This book is written for the construction industry in order to help it provide a better service to clients. Clients are the reason that the industry exists; therefore, clients are always in the industry's thoughts. However, the relationship between clients and the industry is not one of mutual enjoyment. In many cases, there is mutual dislike and distrust, which ends up in acrimonious encounters. Clients see the industry as a problem and often express this feeling in public. The industry sees clients as problems, and although the industry pretends to work for clients, behind closed doors it complains about them.

The fact that construction has its own court system (the Technology and Construction Court is in the Queen's Bench Division of the court service) for dealing with problems pays testimony to the fact that the engagement is not working well. Not all cases that are heard deal with disputes between clients and the industry, but this litigation is the tip of the problem: most disagreements never end up in court, and possibly all construction projects have disagreements between clients and the industry. Although the number of cases in court is reducing, this is a result of new dispute-handling procedures rather than a reduction of disputes. Indeed, such disputes occur everywhere in the world and have occurred throughout history. Thus, our task is momentous. If the problem has not at least been encapsulated so far, is it possible to do it at all? Many have tried, but none has succeeded.

In this chapter, we detail some of the current thinking on clients, considering the way people see their problems, on who clients are and the nature of some of the solutions. This is in no way intended to criticise this other research work, which is well founded and honourable. We then lay out how we intend to present our attempt at grasping the problem and the structure of the book.

1.2   THE NATURE OF THE PROBLEM

The construction industry has been considered problematic for many years. The industry has been the subject of many government reports, which have sought solutions to these problems. Many of these reports are summarised in Murray and Langford (2003), who state:

> The intervention of most of the government reports . . . has been driven by only two groups of powerful clients: in the period 1944–1980 this would have been the powerful government or parastatal clients; while the period 1980–2000 became the era dominated by powerful private clients.
>
> (p. 4)

Government is interested in the construction industry not only because of the influence of powerful clients (including itself) but also because the industry is a national economic driver. Construction creates the buildings and infrastructure that support all other economic activities and so is a necessary component of economic development. In addition, construction is a major employer and user of resources, which creates a stable state. In these respects, then, the government is interested in problems of procurement, continuity of work, fragmentation, productivity and performance (Murray & Langford, 2003). An added complication, however, is that government uses the construction industry for economic stimulus or suppression by instigating (or cancelling) its own projects (Hillebrandt, 1985). With regard to clients, it is fragmentation, productivity and performance of the industry that generate most concern.

In all of the reports summarised by Murray and Langford (2003), what is being emphasised is the failure of the industry to deliver for the client. These failures include:

❏ lack of value for money
❏ budget overruns

❑ non-delivery on time
❑ suspect quality of work
❑ inappropriate buildings
❑ poor client relationships.

The relationship between client and industry is fraught because of its complexity and adversarial nature. Clients see the industry as too complicated to deal with because of its numerous professions and organisations and its convoluted processes and procedures of operation. Furthermore, clients have to experience the industry's internal conflicts that are exposed in projects. Clients suffer the adversarial responses of claims and protective actions, making the experience unpleasant and leading to a continuing lack of trust in the industry's actions and the litigation mentioned above (Banwell, 1964; Egan, 1998; Latham, 1994).

Many reports refer to a better operation of the industry in other countries and to more efficient practices in other industries. There is a clear message that construction in the UK is backward and that its practices need to change (Woudhuysen & Abley, 2004). Each subsequent report, however, laments the fact that the industry has not changed and that the recommendations of earlier reports have not been implemented effectively. This has become the driving force of the reports of the past decade, starting with the Latham Report (Latham, 1994).

Latham (1994) placed 'value for money' at the top of his list of client concerns. His call was for a 30% reduction in the real cost of construction through greater productivity. He identified that this could be achieved by teamwork in the industry. For the first time, however, he also placed a duty on clients to act in ways that would allow this to happen, both in their use of contracts and through their selection of contracting parties on a basis of quality rather than cost only. Latham placed a burden on public-sector clients to work with this idea. The issue of improving value for money in construction was taken up by Atkin and Flanagan (1995), who saw it as part of understanding the client's business needs.

In order to achieve construction customer satisfaction, Egan (1998) proposed five key drivers of change for the construction industry: committed leadership, focus on the customer, integrated processes and teams, quality-driven agenda and a commitment to people. Egan presented the need for benchmarks to be the method of implementing improvements. Benchmark improvements were stated for reductions in cost, time and defects, and measurement of key performance indicators became the driver of change. The methods of improvement that were proposed were process re-engineering and supply chain management, but there was recognition that client leadership was critical and the notion of a best practice client was established. The sequel report, *Accelerating Change* (Strategic Forum, 2002), listed 12 targets and established client satisfaction as a measurable indicator of performance.

The reports before 1994 problematised the industry, and it is only recently that the problem of clients has been acknowledged. In many ways, it has not been possible to criticise clients, as they are paying and should get what they want. It has now been identified, however, that powerful clients can not only get government reports written analysing the problems of the industry but have a duty to develop the industry. This is the theme of the Construction Clients Group, which has established a client's charter and methods of achieving best practice client status. There is a growing realisation that it takes two to tango, and the outcomes of the industry are affected by the actions of the client. Thus, texts not only explain the industry to clients (National Economic Development Office, 1974; Pearce & Bennett, 2003) but also seek to explain to clients how they can act better with the industry for their own benefit (CABE, 2002, 2003; Connaughton & Green, 1996; Construction Industry Board, 1997; Hill, 2000; IOD, 2001; RIBA, 2004).

1.3   THE CATEGORIES OF CLIENTS

In many ways, the problem of clients is contained in the definition of the term 'client'. There is a duality in the term between being a customer paying for goods/services and being a dependent under the protection of another. The term probably was first used as a result of a requirement for legal advice or protection. Nowadays, it is used more broadly and is constituted as a legal entity in a relationship. As Heron (2001) adds, a client is someone who is freely choosing to avail themselves of a service. This wider use includes counselling, social work, hairdressing, beauty therapy, and relationships with bank managers, accountants, management consultants and architects. This root in professional service reinforces the expectation of protection by the client.

The literature on construction clients views clients as the initiators of projects and those that contract with other parties for the supply of construction goods or services (Atkin & Flanagan, 1995). At the end of this relationship, the client has ownership of the outcomes (Miller & Evje, 1999), with legal jurisdiction of the economic advantage (Hillebrandt, 1985). A client can, however, be a representative of the owner and act with delegated authority on the owner's part. This idea of the outcome of the relationship being tied to economic advantage distinguishes construction clients from, for example, counselling clients, where the outcome is individual advantage.

The word 'client' has connotations of an individual as a result of historical and other uses. The historical use has conceptions of a king, nobleman or gentleman instigator of big houses (Hampden-Turner, 1984), which leaves a residual belief in the significance of the individual. In construction, however, it is unlikely that a client will be an individual; most clients are organisations or groups of people. Even when the client is an individual in legal terms, the client's family and associates are part of the decision surrounding the development, as they are beneficiaries or bare a loss. Thus, the client needs to be seen as a plural and involving a set of stakeholders with different viewpoints and different needs (Newcombe, 1994; Rowlinson, 1999; Salisbury, 1998). The perception of clients as unitary entities is constituted in law as the signing party, and this confusion between legal clients and the wider functional clients continuously complicates matters. This is confused further by the representative of the client with delegated authority also being seen as the client.

There are many different categorisations of clients, and Mbachu (2002) distinguishes these as client system-based and client need-based. *Client system-based* categorisation includes:

❑ nature of the organisation
❑ source of the project finance
❑ type of business activity.

Client need-based categorisation includes:

❑ building form types
❑ building use types
❑ ownership type
❑ experience of building.

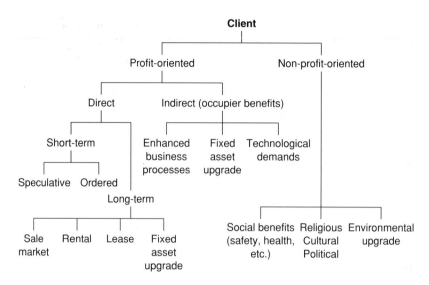

Figure 1.1 Categorisation by organisational form and use.
Source: Adapted from Mbachu (2002).

These describe the starting point of a classification, but there is a choice about the next breakdown at each level, so that the overall classification can unfold in different ways. There is no ideal hierarchical classification system, and problems of interdependence of final categories cause an inadequacy; for example, building use types are linked directly to types of business activity.

Starting with client systems classifications, construction clients can be private individual, private corporate or public sector (Blackmore, 1990). The main distinction is between public and private, and this forms the basis of the Mbachu (2002) breakdown shown in Figure 1.1.

Public-sector (central and local government, nationalised corporations) and semi-public clients collectively form the major segment of construction clients (Hillebrandt, 1985; Egan, 1998) and procure over 40% of construction output by value in most countries (Spencer & Winch, 2002). The significance of public-sector clients is changing in the UK as a result of governmental beliefs that public services can be better provided by the private sector. Thus, the distinction between the public and private sectors is being blurred, although achieving a profit still defines the private sector. There are now private not-for-profit organisations, such as housing associations, that provide social services. The pressures from government on this type of client are to provide greater value for money for the public through numerous initiatives, such as Modernising Construction (NAO, 2001) and Achieving Excellence (OGC, 2003), which have been set in place with this explicit delivery target.

On the private for-profit client side, some clients build for direct profit while others build for self-use (Spencer & Winch, 2002). Most private corporate clients want a building to enhance the value of their other assets, such as to improve a service or to cope with a change in the environment, either because of expansion or because of technological change (Blackmore, 1990). Such desires in buildings, however, potentially also yield greater direct profit if the asset is disposed of, demonstrating the inadequacies of any hierarchical classification system.

Clients' understanding of construction is diverse. Newcombe (1994) and Rowlinson (1999) use three categories: *uninformed (or naive) clients*, who tend to procure projects on either a one-off or a very infrequent basis; *partially informed clients*, who have procured a few projects, often spaced apart; and *well-informed (sophisticated) clients*, who either build on a regular basis or are from the industry.

The majority of construction-industry organisations view clients through their business type or sector categorisation. This is both a marketing strategy and an attempt to provide a better service. We have extended the analysis above to provide us with a categorisation of sectors, as shown in Figure 1.2, and have broken these down into three divisions: public, private and mixed. The public division has two subdivisions: national, whereby services are provided throughout the country, and local, whereby services are resourced and controlled locally. Clearly, the advent of regional government may add other tiers to this. The private division has two subdivisions: industrial and service. These categories are used in economic categorisation and agencies such as Business Link. The final division is more complicated and involves a degree of mix of public and private enterprise. The divisions are not-for-profit and private regulated. The not-for-profit company is a privately constituted organisation that provides a public service but channels its surpluses back into the service. A private regulated organisation undertakes a public service, often with a monopoly, but it is allowed to make a profit. The regulation defines and monitors the service that is offered and tries to limit the profit through pricing controls.

This list is not comprehensive and there may be anomalies. Some of the sectors mentioned can appear on both the public and the private

Local	National	Industry	Service
Schools	Defence	Agriculture and	Business services
Social	Government	fishing	Creative services
services	Highways	Biotechnology,	and media
Leisure	Environment	medical and	Education
Police	Prisons	chemical	Financial services
Waste	Courts	Construction and	Health and social
Transport	Taxation	building services	work
Public	Intelligence	Energy and water	Hospitality and
administration	Regulators	Manufacturing and	leisure
Fire		engineering	IT and telecoms
Roads		Mining	Personal services
Libraries		Publishing, printing	Professional services
Housing		and packaging	Real estate
			Retail and wholesale
			Transport, storage
			and distribution
			Travel and tourism

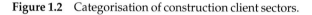

Mixed

Not-for-profit	Private regulated
Rail	Water
Housing associations	Airports
Universities	Ports
Religious	Electricity
TV and radio (BBC)	Gas
Nuclear power	Telecommunications

Figure 1.2   Categorisation of construction client sectors.

sides, for example education and healthcare. In most of these cases, however, it is the public sector that dominates. The only exception is social housing: most social housing is now provided by not-for-profit organisations, with the public sector providing only specialist accommodation.

Moving on, we can start to look at specific clients. The scale of client business is clearly an important issue, as larger organisations procure a larger amount of building on a more regular basis. Both *Building* magazine and *Construction News* provide regular lists of the top 50 clients by expenditure. The list for May 2004 to April 2005 is shown in Table 1.1.

This list is dominated by the public sector, with the top two and five of the top ten. It is clear that one or two projects can dominate the statistics, which means that in the following year this work may not be ongoing. As Hewes (2003) points out, infrastructure clients will

Table 1.1 Top 50 UK clients by expenditure from May 2004 to April 2005.

Rank	Firm	No.	Total (£m)
1*	Department of Health	239	2,637.3
2*	Department for Transport	55	1,135.6
3	News International	6	678.6
4	Land Securities	32	433.9
5*	Ministry of Defence	55	289.1
6	English Partnerships	30	278.8
7*	Scottish Executive	41	275.6
8	ASDA	22	269.0
9	National Grid Transco	11	268.0
10*	Leeds City Council	36	259.1
11	Berkeley	14	224.9
12	Westfield	3	203.0
13	CTP	7	189.5
14*	Liverpool City Council	7	119.8
15*	Sunderland City Council	16	177.0
16	Tesco	43	175.6
17	Wilson Bowden	19	164.2
18*	Home Office	61	157.3
19	Union Railways	6	154.1
20*	Exeter City Council	3	142.5
21	British Land	3	138.0
22*	Network Rail	24	126.3
23	Wm Morrison	30	125.4
24	ProLogis	8	124.6
25	Scottish Hydro Electric	8	120.3
26	Allied London	2	120.0
27	Rosemound	12	118.2
28*	University of Manchester	11	116.2
29*	Southbank Centre	3	113.0
30*	Solihull MBC	4	112.7
31*	Renfrewshire Council	3	106.5
32*	Rhondda Cynon Taff CBC	5	102.5
33*	Bolton Borough Council	3	101.7
34*	Cornwall County Council	12	100.4
35	Rolls Royce	4	99.8
36*	West Sussex County Council	20	99.1
37	J Sainsbury	19	98.9
38*	National Assembly for Wales	10	85.9
39*	Telford & Wrekin Council	6	81.1
40	Scottish Widows	3	80.0
41	Barratt	7	79.3
42*	DEFRA	9	79.0
43	London Bridge	2	77.0
44*	University West of England	2	75.6
45	Gladedale	9	75.5
46	London & Amsterdam	2	75.0
47*	Doncaster MBC	10	74.3
48	Redrow	5	73.4
49*	Dept Regional Development	34	71.7
50*	Wrexham CBC	4	71.3

Source: *Construction News*, 19 May 2005.
*Public sector.

dominate because this sector lends itself to a small number of large firms. In other sectors, there is a large number of smaller clients or, as in food retailing, four large clients and the remainder very small clients.

1.4   WHAT CLIENTS WANT

There are two problems for clients: expressing what they want and getting this delivered. We deal first with what clients want before handling the delivery processes in the next section.

Groak (1992) declares that the orthodoxy of social demand for buildings is that 'people have always wanted much the same thing: firmness, commodity and delight'. This reference to Vitruvius, the 1 BC Roman author of *Ten Books on Architecture*, ably sums up a large amount of literature on what buildings are required to be: firmness being technically sound, commodity being of use and created with minimum resources, and delight having a presence in the environment. Boyd and Danks (2000) expand on this, suggesting that there are four economies to be negotiated in building: the functional economy, the financial economy, the social economy and the symbolic economy. Each of these provides us with a list of different needs, not all of which can be provided and some of which are in conflict; however, the resolution is undertaken as a negotiation about meaning and significance. Boyd and Danks use the term 'economy' to identify the subject area of the set of negotiated transactions involved in the task of building. These are shown in Figure 1.3; we have indicated that functional and symbolic needs are experiential, while financial and social needs are abstract. This helps us to understand the difficulties in achieving a resolution, as the items are so different and involve different forms of thinking.

The functional and financial economies are well understood, but the social and symbolic economies require explanation. Buildings do not only perform a technical function; nor are they simply aesthetic artefacts. As Zukin (1996) states, 'Framing architecture is not primarily a visual problem; it is a problem of social power'. Thus, we see on the one hand design as a commodity to be packaged and sold for mass consumption and on the other hand an exclusive statement to declare individual difference and importance to anyone else. The social economy involves a long-term integration of a building into an environment. This has ranged from a belief that buildings can be part of an engineering of society to the upholding of a wider social identity in

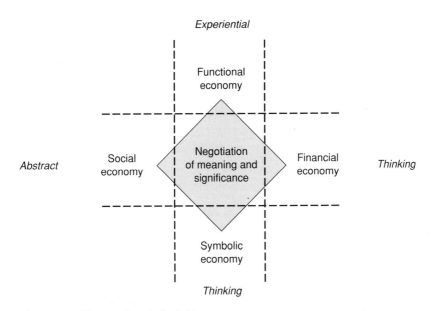

Figure 1.3   Economies of a building.
Adapted from Boyd and Danks (2000).

towns and cities through the planning system and, recently, of providing for sustainability.

This identifies the problem of the task of establishing clients' needs. Although some clients' needs are expressible to the construction providers, other needs remain latent, concealed either consciously or unconsciously (Kamara *et al.*, 2002). Runyon (1980) supports the existence of concealed needs and objectives in the mind of the 'consumer', observing: 'Often, a consumer's goals for making purchase decisions are complex, not easily inferred from direct observations primarily because the respondent may be unwillingly to release such information.' Such hidden objectives may include the provision of status and prestige for the client. Bennett (1985) explains that needs and objectives may be concealed because they may have social, cultural, political or religious connotations. Thus, latent needs may not be observed easily and cannot be accessed merely by asking; and yet clients assess overall satisfaction with the procurement outcomes or the services provided by taking into consideration the extent to which these latent needs are met (Green & Lenard, 1999; Salisbury, 1998; Turner, 1997).

These problems are addressed by briefing. Briefing is a dialogue between the client and the construction professionals, normally the architect (Gameson, 1991; Loe, 2000), where the client's aspirations, desires and needs are expressed and presented in a written form called the 'brief' (BSI, 2002; Construction Industry Board, 1997). Early views on briefs considered it essential for the client to have a clear view of what the facilities should achieve and why they are necessary before starting any construction project. Briefs then needed to be clear and to be fixed early in the project in order to enable the construction team to undertake its job (Kelly *et al.*, 1992). A number of problems were found with this conventional approach: there was a continued lack of satisfaction from construction clients (Green & Simister, 1999), and the effectiveness of briefings has remained problematic (Shen *et al.*, 2004). Sometimes the briefing function was not accorded due priority by clients and thus was not carried out properly. There were calls for large client organisations to appoint a construction champion who would speak for the client. This idea, however, contradicted the aware-ness that, with bigger organisations, there were many relevant voices and different needs in the project from these. Most often, there was a distinction between the ownership and occupation of buildings, cloud-ing the identity of the client (Newcombe, 2003). The opinions and needs of users are important in briefing, but it was realised that it was impossible to satisfy all users (Blackmore, 1990). Indeed, although the brief could capture the opinions of all stakeholders, it could not help to negotiate compromises. In addition, as projects take extended periods of time, personnel and stakeholders can change and these can challenge the brief in order to have their needs met. Further, clients become more aware with time and realise or discover new needs or opportunities as their projects proceed (Beijder, 1991). Clients' object-ives thus change as they engage with designers and other consultants to determine their real and explicit needs (Powell, 1991). The irony then is that clients attach a high expectation to some of the needs that are discovered after the brief is formulated, which pressurises the delivery of the construction product.

The changing and modern face of construction delivery has demanded a move away from traditional briefing to more elaborate approaches that will decipher the requirements of clients and expectations of stakeholders. Atkin and Flanagan (1995) suggested that a modern-day briefing would include strategic analysis, client analysis, facilities analysis, statement of need, confirmation of need, functional brief, concept design and scheme design. In a sequel, Smith *et al.* (2001) emphasised the need for strategic clients' needs analysis.

Green (1994) endeavoured to account for the conflicting and transient aspirations of the project stakeholders using value management. Green and Simister (1999) trialled the viability of three soft operational research methodologies as enablers of strategic briefing, i.e. 'soft systems methodology' (Checkland, 1981, 1989), 'strategic choice' (Friend & Hickling, 1987) and 'strategic options development and analysis' (Eden, 1989). This proved extremely successful for clients in reaching a consensus brief rather than attempting to optimise the brief. Green (1996) sought an even better approach to understanding clients' requirements by seeing them as social systems that can be mapped against or viewed through the lenses of Morgan's (1986) eight metaphors, i.e. goal-seeking machine, biological organism, intelligence, culture, politics, psychic prison, 'flux and transformation' and domination. Depending on the culture in an organisation, one or more metaphors will be dominant and influential on the client's requirements and procedures.

McGregor and Then (1999), using a facilities management perspective, saw that what was required was a more detailed awareness of the client's business. This involved determining the position of buildings in the business and how the client made use of space. Barrett and Stanley (1999) saw the wider processes of engagement with the industry as being important and also redefined the client's attempts at conceptualising the precise nature of a building and its uses as a journey from uncertainty to certainty and from aspiration to delight. Barrett and Stanley offered a more structured approach to this whole process, involving five key solutions (Figure 1.4). The act of briefing in which these five solutions are evolved is an interactive process that runs concurrently with the construction project and is not a single rational event.

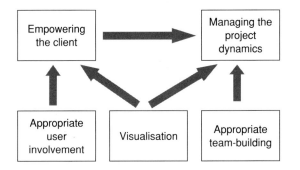

Figure 1.4   Modern client briefing processes.
Source: Adapted from Barrett and Stanley (1999).

The issues with the foregoing are that the client is empowered to play its role in the processes in an effective way but also the client needs to feel that it is in control. However, the changes during the project that constitute Barrett and Stanley's (1999) journey from uncertainty to certainty need to be managed so that they are realistic and integrated. There needs to be appropriate user involvement from within the client to ensure that the wider client is satisfied. There also needs to be a wider involvement from the industry team in the process, and this team needs to learn how to work together. Finally, Barrett and Stanley (1999) believe that new techniques of visualisation can be at the centre of improving communications and of understanding what is really required.

Given the enormous effort needed for developing detailed briefs, especially in complex projects, the role of the briefing consultant has become a recurring theme (BSI, 2002; Hyams, 2001). The briefing consultant has an understanding of clients' businesses and construction and can drive the brief formulation process alongside the project process. The consultant is an independent adviser, playing a similar role to a financial adviser, and is there to help clients decide on their needs and how to achieve them (Myers, 2004).

1.5   A PROBLEM OF DELIVERY

In order to have a building built, the client needs to interact with the industry. This requires engaging with a diverse group of professionals and businesses and entering into contracts. This process is normally called 'procurement' and has become the subject of development within the industry. Procurement is the framework within which construction is brought about, acquired or obtained. In the traditional form of procurement, the client decides what is wanted; appoints and briefs a designer (architect or engineer), who designs the building and who also informs the selection of a contractor, who will deliver what has been designed. In this, the client has to face the problems endemic in the construction industry.

The economics of the industry have caused it to have a complex fragmented structure (Ball, 1988; Hillebrandt, 1985). The industry has to face varying demands for its work, which has caused it to fragment itself through subcontracting (Smith *et al.*, 2004). This means that a different group undertakes every project, and that individual organisations may be undertaking any number of jobs at once. Rather than manage the technical aspects of building, most management is of contracts between organisations. The processes of competitive tendering place contractors in a high level of competition, which encourages opportunistic behaviour through either overpricing (for thriving contractors who may not want to turn down a chance to bid in view of future opportunities) or underpricing (on the part of anxious contractors who are eager to gain work).

Negotiation can put either the client or the contractor in an opportunistic position, depending on the situation. For instance, the contractor involved in a certain negotiation could be in a position of monopoly because other competitors are reluctant to take on work. Given that transactions costs are high in construction procurement (Winch, 2002), the opportunity costs of a client walking away from negotiations are equally high, thus putting the contractor in an exploitative position. It is then left to that contractor to act either opportunistically or altruistically (Winch, 2002).

The Tavistock Institute viewed the construction industry as consisting of five subsystems: operations, resource control, directive function, adaptive functions, and social and personal relations (Crichton, 1966). Because of this, project decisions were not discrete but were interdependent and beset by several uncertainties. There were several interfaces and several potential conflicts of interests; if these were not resolved or managed, then a successful outcome was jeopardised. Cherns and Bryant (1984) saw each project team as a temporary multiorganisation that depended on the nature of the project, the client's choice of professionals, and the form of procurement. Each project team member was seen to come from an organisation that had its own internal conflicts that contributed to the risks. Each project had its own complexities that were changing in time, and the project was difficult to manage. For a client undertaking construction, this was frustrating and sometimes expensive.

Inexperienced clients believed that they were safe in the hands of consultants who were working for them, but these consultants were also fragmented around professional roles backed up by professional institutions with a vested interest in maintaining the system (Bowley, 1966; Masterman, 2002). These professionals were trained to see the world in different ways; when they had to work together, this often resulted in conflict. In a similar way to contractors, consultants could act individualistically. Designers are sometimes reluctant to change their proposals as clients' needs change; instead, designers tend to hold on to their initial solutions (Hyams, 2001). The myth is that architects go for creative designs at the expense of clients' needs. There are also design management problems that concern working with the brief and coordinating numerous technical components (Gray & Hughes, 2000). Thus, there is a tendency not to design completely and to leave decisions to be expedited on site; these are occurring within a climate of competition on fees driven by the client's desire for lower costs.

Together these create a series of management problems for the client: the problem of engaging the industry as a whole, the problem of telling the industry what is wanted, the problem of ensuring the design addresses what is wanted, the problem of paying for this design, the problem of employing a contractor in order to make the design real, the problem of facing the changes in conception, and the problem of sorting it all out at the end. The rise of project management was meant to alleviate these problems (Walker, 2002). The project manager's task was to integrate the various parties and ensure that progress took place to the client's satisfaction. At best, this has worked; however, there are many conceptions of project management, some of which simply involve administrating the system (CIOB, 2002).

Given the problems of determining a fixed brief and the problems of delivery, it is not surprising that clients have often experienced problematic building. Experienced clients know what they want and have taken decisive measures to make sure they achieve it (Latham, 1994). These clients, such as fast-food outlets, retail stores and petrol stations, have a continuity of demand that allows them to standardise the product, reduce the need for briefing and learn from previous projects. Other clients have to face the problematic industry with the potential of failure and anger at the cost. Thus, we return to the reports on the construction industry and the initiatives to make the industry work. No one can deny that there is a problem, but no one is really sure what the answer is.

1.6   STRUCTURE OF THIS BOOK

This book can be seen in three parts: a theoretical analysis of clients, a study of six client sectors, and a toolkit for the industry to engage better with clients.

The theoretical analysis is provided in Chapters 2–4. Chapter 2 outlines the model that will be used throughout the book. This is a complicated model in order to try to encompass all situations. The model is related to some fundamental observations about clients and industry engagement. It is a system dynamics model that works on three areas in which the client needs to be satisfied: the building, the organisation and the organisation's people. Each of these areas involves two dimensions: a means and ends dimension, which concerns change and experiences; and a knowledge and processes dimension, which contains how the client thinks and makes decisions.

Chapter 3 considers the theory behind the knowledge and processes dimension. This is based on observing the client at rest, i.e. when not building. The client's organisational objective is to maintain stability. The significance is that there are a number of stresses on the client even at rest, including external pressures, organisational differentiation, and the emotional drives of the organisation's people. These stressors induce unknowns within the client organisation. They are coped with in normal operation through informal processes that are also in the unknown.

Chapter 4 considers the theory behind the means and ends dimension. This is based on the client's experience of change. Building involves both project change and organisational change, and thus the desire for stability is upset. As well as this upset, the unknown informal coping mechanisms, which work when the client is at rest, are no longer effective and the organisation exaggerates gaps and contradictions within it and exposes the emotional consequences. This is made worse by the change processes involving unformed unknowns that come into existence only during the change. The success of the project and organisational change is evaluated against gaps between rational plans and reality and between aspirations and reality.

In order to make effective use of this model to improve the engagement between the industry and clients, it is necessary to have a greater understanding of the client's business, particularly the client's knowledge and processes and the client's approach to handling unknowns. A description and analysis of six client sectors is provided in Chapters 5–10. These sectors are supermarkets, property developers, NHS trusts, government departments, airports and housing associations. The selection of these was based partly on the top 50 clients shown in Table 1.1. These are all significant clients and procure significant amounts of construction. The six sectors are split between public (NHS trusts, government), private (supermarkets, developers) and mixed (airports, housing associations). We believe that some will be expanding their construction in the next 10 years, namely airports and NHS trusts, and thus we are preparing the industry to address future expanding sectors. Others are concerned with a continuity of work, i.e. property developers and supermarkets. The government will always be a major client because of its responsibility for providing infrastructure, and its approach to procurement is changing with a greater use of public–private partnerships. Indeed, we have had to cope with a number of changes in the sectors while writing this book. In particular, NHS trusts seem to have gone through three metamorphoses. We cannot bring these chapters up to date, and therefore we have not tried, but we state when the picture was taken. In some ways, the advantage of this approach is that it establishes longer-term values rather than merely presenting structures at any one instant.

The toolkit for engagement is presented in Chapter 11. This is based on the model and the use of client sector analyses. We believe it is the industry's duty to learn to work with clients better. The industry must help them not only with building but also with organisational change. The toolkit establishes a requirement for a new skill of process consultation, which aims to help the client deal with the unknowns within their system and in the project. We believe that this will allow current formal approaches of briefing, team-building, value management, project management, risk management and partnering to work more effectively.

Chapter 12 looks at the methodology used within this book and considers the inadequacies of the model and the approach, as it is important that we justify the hows and whys of the approach.

Finally, an appendix lists our interviewees. Their cooperation and knowledge are priceless, and the book owes much to them.

1.7 A CONCLUDING REMARK

In construction, the need to improve is clear. Clients need better value from their projects, and construction companies need reasonable profits in order to assure their long-term future. The impetus for change must come from major clients. Egan's taskforce arrived at this conclusion by comparing construction with other industries that have increased their efficiency significantly, e.g. automotive, aerospace and steel manufacturing. The taskforce observed that the construction industry tends to think more about the next employer in the contractual chain than about the current customer. Construction companies carry out little systematic research on what the end user actually wants; neither do they seek to raise customers' aspirations and educate them to become more discerning. Meanwhile, the aforementioned comparative industries hold the client as king. Construction clients have increasing expectations on the industry of today and have an acute longing for greater satisfaction. This book tries to make this happen.

REFERENCES

Atkin, B. and Flanagan, R. (1995) *Improving Value for Money in Construction.* London: Royal Institution of Chartered Surveyors.

Ball, M. (1988) *Rebuilding Construction.* London: Routledge.

Banwell, H. (1964) The Placing and Management of Building Contracts. The Banwell Report. London: HMSO.

Barrett, P. and Stanley, C. (1999) *Better Construction Briefing.* Oxford: Blackwell Science.

Beijder, E. (1991) From client's brief to end of use: the pursuit of quality. In *Practice Management: New Perspectives for the Construction Professional,* eds P. Barrett and R. Males. London: Spon, pp. 125–136.

Bennett, J. (1985) *Construction Project Management.* London: Butterworths.

Blackmore, C. (1990) *The Client's Tale: The Role of the Client in Building Buildings.* London: Royal Institute of British Architects.

Bowley, M. (1966) *The British Building Industry: Four Studies in Response and Resistance to Change.* Cambridge: Cambridge University Press.

Boyd, D. and Danks, S. (2000) An analysis of the architectural practice in the construction industry. In *Proceedings: 16th Annual Conference of the Association of Researchers in Construction Management,* ed. A. Akintoye. Reading: Association of Researchers in Construction Management, pp. 693–702.

BSI (2002) *Performance Standards in Building: Checklist for Briefing – Contents of Brief for Building Design.* London: British Standards Institute.

CABE (2002) *Client Guide for Arts Capital Programme Projects.* London: Commission for Architecture and the Built Environment.

CABE (2003) *Creating Excellent Buildings: A Guide for Clients.* London: Commission for Architecture and the Built Environment.

Checkland, P. (1981) *Systems Thinking, Systems Practice.* Chichester: John Wiley & Sons.

Checkland, P. (1989) Soft systems methodology. In *Rational Analysis for a Problematic World: Problem Structuring Techniques for Complexity, Uncertainty and Conflict,* ed. J. Rosenhead. Chichester: John Wiley & Sons, pp. 71–100.

Cherns, A. and Bryant, D. (1984) Studying the client's role in construction management. *Construction Management and Economics,* **2**, 177–184.

CIOB (2002) *Code of Practice for Project Management for Construction and Development.* Oxford: Blackwell.

Connaughton, J. and Green, S. (1996) *Value Management in Construction: A Client's Guide*. London: Construction Industry Research and Information Association.

Construction Industry Board (1997) *Briefing the Team: A Guide to Better Briefing for Clients*. London: Thomas Telford.

Crichton, C. (1966) *Interdependence and Uncertainty: A Study of the Building Industry*. London: Tavistock Publications.

Eden, C. (1989) Using cognitive mapping for strategic options development and analysis (SODA). In *Rational Analysis for a Problematic World: Problem Structuring Techniques for Complexity, Uncertainty and Conflict*, ed. J. Rosenhead. Chichester: John Wiley & Sons, pp. 21–42.

Egan, J. (1998) *Rethinking Construction: A Report of the Construction Task Force to the Deputy Prime Minister, John Prescott, on the Scope for Improving the Quality and Efficiency of UK Construction*. London: Office of the Deputy Prime Minister.

Friend, J.K. and Hickling, A. (1987) *Planning Under Pressure: The Strategic Choice Approach*. Oxford: Pergamon Press.

Gameson, R.N. (1991) Clients and professionals: the interface. In *Practice Management: New Perspectives for the Construction Professional*, eds P. Barrett and R. Males. London: Spon, pp. 165–174.

Gray, C. and Hughes, W.P. (2000) *Building Design Management*. London: Butterworth-Heinemann.

Green, S.D. (1994) Beyond value engineering: SMART value management for building projects. *International Journal of Project Management*, **12**(1), 49–56.

Green, S.D. (1996) A metaphorical analysis of client organizations and the briefing process. *Construction Management and Economics*, **14**(2), 155–164.

Green, S.D. and Lenard, D. (1999) Organising the project procurement process. In *Procurement Systems: A Guide to Best Practice in Construction*, eds S. Rowlinson and P. McDermott. London: Spon, pp. 57–82.

Green, S.D. and Simister, S.J. (1999) Modelling client business processes as an aid to strategic briefing. *Construction Management and Economics*, **17**(1), 63–76.

Groak, S. (1992) *The Idea of Building*. London: Spon.

Hampden-Turner, C. (1984) *Gentlemen and Tradesmen*. London: Routledge and Kegan Paul.

Heron, J. (2001) *Helping the Client: A Creative Practical Guide*, 5th edn. London: Sage.

Hewes, M. (2003) Where the smart money is. *Building Magazine*, 28 February.

Hill, R.M. (2000) *Better Building: Integrating the Supply Chain – A Guide for Clients and their Consultants*. Watford: CRC.

Hillebrandt, P. (1985) *Economic Theory and the Construction Industry*, 2nd edn. Basingstoke: Macmillan.

Hyams, D. (2001) *Construction Companion to Briefing*. London: Royal Institute of British Architects.

IOD (2001) *Buildings that Work for Your Business*. London: Institute of Directors.

Kamara, J.M., Anumba, C.J. and Ovbuomwan, N.F.O. (2002) *Capturing Client Requirements in Construction Projects*. London: Thomas Telford.

Kelly, J., McPherson, S. and Male, S. (1992) *The Briefing Process: A Review and Critique*. London: Royal Institution of Chartered Surveyors.

Latham, M. (1994) *Constructing the Team: Final Report of the Government/ Industry Review of Procurement and Contractual Arrangements in the UK Construction Industry*. London: The Stationery Office.

Loe, E. (2000) *The Value of Architecture: Context and Current Thinking*. London: Royal Institute of British Architects Future Studies.

Masterman, J.W.E. (2002) *An Introduction to Building Procurement Systems*. London: Spon.

Mbachu, J.I. (2002) Modelling Client Needs and Satisfaction in the Built Environment. Unpublished PhD dissertation. Port Elizabeth: University of Port Elizabeth.

McGregor, W. and Then, S.-S. (1999) *Facilities Management and the Business of Space*. London: Arnold.

Miller, J.B. and Evje, R.H. (1999) The practical application of delivery methods to project portfolios. *Construction Management and Economics*, **17**, 669–677.

Morgan, G. (1986) *Images of Organization*. London: Sage.

Murray, M. and Langford, D. (eds) (2003) *Construction Reports 1944–98*. Oxford: Blackwell.

Myers, D. (2004) *Construction Economics: A New Approach*. London: Spon.

NAO (2001) *Modernising Construction*. London: National Audit Office.

National Economic Development Office (1974) *Before You Build: What a Client Needs to Know about the Construction Industry*. London: HMSO.

Newcombe, R. (1994) Procurement path: a proper paradigm. In *East Meets West: Proceedings of CIB W92 Procurement Systems Symposium*, ed. S.M. Rowlinson. Hong Kong: University of Hong Kong, pp. 245–250.

Newcombe, R. (2003) From client to project stakeholders: a stakeholder mapping approach. *Construction Management and Economics*, **21**, 841–848.

OGC (2003) *Building on Success: The Future Strategy for Achieving Excellence in Construction*. London: Office of Government Commerce.

Pearce, S. and Bennett, J. (2003) *A Client Guide: How to Use the Construction Industry Successfully*. Ascot: Chartered Institute of Building.

Powell, J.A. (1991) Clients, designers and contractors: the harmony of able design teams. In *Practice Management: New Perspectives for the Construction Professional*, eds P. Barrett and R. Males. London: Spon, pp. 137–148.

Pries, F., Doree, A., Van Der Veen, B. and Vrijhoef, R. (2004) The role of leaders' paradigm in construction industry change. *Construction Management and Economics*, **22**(1), 7–10.

RIBA (2004) *A Client's Guide to Engaging an Architect*. London: Royal Institute of British Architects.

Rowlinson, S. (1999) The selection criteria. In *Procurement Systems: A Guide to Best Practice in Construction*, eds S.M. Rowlinson and P. McDermott. London: Spon, pp. 276–279.

Runyon, K.E. (1980) *Consumer Behaviour and the Practice of Marketing*, 2nd edn. Toronto: Charles E. Merrill Publishing Company.

Salisbury, F. (1998) *Briefing Your Architect*. Oxford: Architectural Press.

Shen, Q., Li, H., Chung, J. and Hui, P.-Y. (2004) A framework for identification and representation of client requirements in the briefing process. *Construction Management and Economics*, **22**(2), 213–221.

Smith, J., Love, P.E.D. and Wyatt, R. (2001) To build or not to build? Assessing the strategic needs of construction industry clients and their stakeholders. *Structural Survey*, **19**(2), 121–132.

Smith, J., O'Keeffe, N., Georgiou, J. and Love, P.E.D. (2004) Procurement of construction facilities: a case study of design management within a design and construct organisation. *Facilities*, **22**(1/2), 26–34.

Spencer, N. and Winch, G.M. (2002) *How Buildings Add Value for Clients*. London: Thomas Telford.

Strategic Forum (2002) *Accelerating Change*. London: Rethinking Construction.

Turner, A. (1997) *Building Procurement*, 2nd edn. London: Macmillan.

Walker, A. (2002) *Project Management in Construction*, 4th edn. Oxford: Blackwell Science.

Winch, G.M. (2002) *Managing Construction Projects*. Oxford: Blackwell.

Woudhuysen, J. and Abley, I. (2004) *Why is Construction so Backward?* Chichester: John Wiley & Sons.

Zukin, S. (1996) Power in the symbolic economy. In *Reflections in Architecture: Practices in the Nineties*, ed. W.S. Saunders. New York: Princeton Architectural Press.

2 A Model of Clients

2.1 INTRODUCTION

Chapter 1 set up our challenge for this book, namely that clients are dissatisfied with the industry and the industry needs to learn to deal with this in order to create satisfied clients. This book is seeking a better engagement between the construction industry and its clients. Such engagement can never be perfect, but we should be able to make it better on more occasions. The approach we are adopting is to seek an understanding of clients in general and then to use this to consider specific clients. This distinction is important: in reality, every client is specific and no general theory or understanding can be sufficient. We are conscious that as an industry, constructors are prone to acting from experience (Boyd & Wild, 1994). There is a tendency to eschew understanding and to move directly to action. However, in order for this action to address specific clients, it requires a specific approach for each client, which needs to be developed from a more general understanding. In the end, we will use this understanding to develop a model of engagement that draws out the problems and identifies what needs to be managed. This understanding will be used to analyse clients in a number of sectors, which will move the manner in which they are understood towards the specific. After that, the specific can be dealt with by suggesting what needs to be found out in order to engage with individual clients fully and successfully. As we mentioned in Chapter 1, we will provide a toolkit for engaging in a better manner with specific clients.

The model is complex, in order to deal with the complexity of the building situation and the diversity of clients. It is a model of models about clients and interactions. Thus, there are multiple layers of meaning within the model, which need to be explained separately. The approach we have adopted is first to reveal the complete model and then to explore the details and sub-models within it and the justification for them. Doing this helps one to appreciate the whole before the parts, which mirrors one of the strategic aspects of our approach for the industry, i.e. to understand broadly about a client before launching into providing a 'solution' to the client's detailed needs.

There is an interesting extension of this idea of 'solutions', which is current in the management and organisational world. We are no longer doers, i.e. builders, engineers and architects; we are 'solution providers' (Dawson, 2000; Drucker, 2002). This more abstract concept of business and service is necessary because of the complexity of the world and the fact that clients cannot or do not want to waste resources in order to understand everything. Thus, clients require not a fixed off-the-shelf product but something that adds value to their businesses. What the industry offers, then, is not a building or a design but something that interacts with the problem or opportunity within the client's thinking. This directs us to a more abstract concept of business and service as knowledge processing rather than exploitation or creation of fixed capital (Drucker, 1998). Thus, we need to start working with 'the mind of the organization' (Heirs & Pehrson, 1982) and helping the client manage its knowledge (Dawson, 2000), especially where this knowledge is incomplete or in conflict, and this will be our route.

The thing about models is that they are models: they are not the thing they represent (Flood & Carson, 1993). Models are not perfect, are not complete and are only partial representations of what is possible. In this, they are a way of looking at and exploring the world. Beware! These types of model do not have the same character as a model for calculating the bending of a steel beam, which has a universal correctness in most situations upon which everyone can agree on and which works. The difficulty and uselessness in developing a similar deterministic model for the engagement with the client is important and is one of the understandings required to engage better. Clients are not like steel beams, but we perceive that the industry may treat clients as if they were, which might be one of the problems. Some people like working with models, but some do not; this polarisation of personality makes life hard. It makes life hard for us writing this book, as there are various different people in the readership. More importantly, it makes life hard for the industry's engagement with clients: what happens if we and they are different people? The model has to account for this perceptual diversity.

Having made these important points about models generally, why use multiple models? One might be annoyed by the idea of several models, hoping instead for a single unified theory. Although the latter might appear attractive, it fails to mirror adequately the approach we are suggesting. First, each model focuses on something different. This is good, as it allows us to not lose track of some things in relation to others. Second, it simplifies our ability to understand the models, and the explanation of them relates to their usefulness. Third, some models are more universal than others, so that a model that requires to be developed in relation to specific clients can be addressed and explained in a different way to a model that is generally applicable. We will start with a general view and then look in detail at some of the other models. Fourth, concepts that appear in a model have different meanings in different models and this could get lost in a single model. However, we do realise that extra effort is required to successfully bring together models on construction clients in order to create a whole theory; thus, it is important that we start with a clear statement of clients and buildings. This statement we call the basic thesis.

2.2   THE BASIC THESIS

Our understanding of clients and their interaction with the industry is based on the following thesis:

> Clients see and experience building differently from the industry. What building involves is change in the client organisation. Buildings are large, and therefore the change can be large. Change puts pressure on the client organisation and its people, as it involves actions and experiences outside the norm and this exposes gaps and contradictions in this norm. In particular, the different objectives of the divisions within the client organisation induce conflict. This situation is made worse by the fact that building involves unknowns that are yet unformed, coming into existence only during the process of building, which limits the ability to deliver plans to pre-established expectations. The organisational change in the client and the experience of unknowns during building induces emotions in the client, which feed back to affect the situation, both the process of building and the organisational change. The industry is used to these changes and unknowns; the client organisation, however, is not.

In this thesis there are a number of fundamental awarenesses that drive the model, which must be emphasised:

❑ Clients see building differently from the industry.
❑ Clients are not unitary.
❑ Building involves organisational change in the client.
❑ Change exposes gaps and contradictions in the client.
❑ Building involves unknowns that are not yet formed.
❑ Building is emotional for the client.

We are acutely aware that the consequences of this are dependent on the structure, processes and historical operation of the industry. This will be dealt with separately in Chapter 4.

2.3   A MODEL OF CLIENTS

The basic structure of the model is shown in Figure 2.1. This is a system dynamics model (Flood & Carson, 1993; Sherwood, 2002) based on constraints external to the situation, as in contingency theory (Huczynski & Buchanan, 2001), from the constitution of the system (Goldratt *et al.*, 2000) and of individual and group psychodynamics (French & Vince, 1999). The model involves three areas in which the client needs to be satisfied: in the building, as an organisation and as people. Each of these involves two dimensions: a means-and-ends dimension that concerns change and time (i.e. what the client experiences), and a knowledge-and-processes dimension that involves conception, communications and decision-making (i.e. how the client sees things, makes decisions and tells people about this). Of particular importance here is what happens when the client does not know, as this challenges the client's normal organisational processes concerned with effectiveness of stable operation. Outside this is the set of external factors that impinge on the client and the project. The industry is part of the external world, which enters this world of the client.

We will call the key elements of the model (i.e. buildings, organisations and people) the three achievements (of satisfaction). Buildings are the normal achievement of the industry, which previously was all that was required. Hidden in our idea of building, however, is the second achievement area – the organisation – as there is an organisational reason for the client wanting the building, i.e. the business or

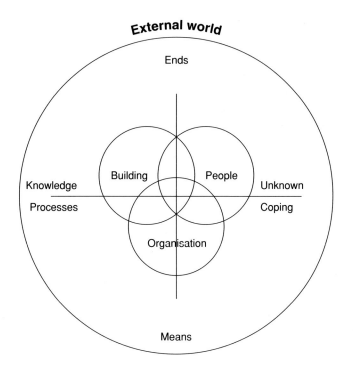

Figure 2.1 Basic model for understanding clients.

service objective, and an organisational means of financing it. Organisational satisfaction is often the most important area. The last of our three achievement areas concerns people: the client is not only an organisation but also a group of individual people with needs to be satisfied and emotions to be expressed. These needs and expectations may be different, but also they interact to create new ideas and actions, which induce further emotions.

In order to work towards these achievements, we need to understand the two dimensions that interact to deliver the satisfaction of the achievements. First, as building involves both physical and organisational change, and because of the interaction between our achievements, then it is not just the end achievement that is important – it is the means or way in which this end achievement is achieved over time that is important. This is clearly more important for the organisation and the people achievements, although the integrity of the final building is also a function of its means of design and construction. The fact that the means of achievement are important determines that every action in the means (intended or unintended) may jeopardise achieving the end satisfaction, depending on how this is seen by the client. The way the client sees (i.e. perceives, thinks and makes decisions) is contained in the other dimension in the model: knowledge and processes. Also the end may change during the process, as the end is determined by the means. This effectively involves a change in objectives, and this may jeopardise the end satisfaction, depending again on how this is seen by the client.

In organisational terms, the client's knowledge and processes determine how the client sees and experiences the world, and this is critical in determining how the client experiences satisfaction. These knowledge and processes also determine how a particular building project is instigated, and this interacts with the means and ends. However, it is not just what is known that is important. At the other end of this dimension, there is what is unknown. As the end is in the future, then the process of change (physical and organisational) and the involvement of people mean that the end is not known completely. Thus, how the client deals with unknowns and uncertainty becomes a critical part of building and organisational change.

Organisational knowledge and processes tend to be constructed around the normal activity of the business or service in order to optimise the efficiency and/or effectiveness of the business. During change, this knowledge and these processes are unlikely to be effective, as the conditions of their constitution are no longer applicable. The organisation is then in a coping dilemma as to whether to force its processes to work in order to maintain the integrity of the organisation and yet not handle the uncertainty well, or whether to loosen its process and structure in order to handle the situation – but this entails losing the coherence of the organisation and the efficiency of its normal operation. This dilemma puts pressure on the organisation, which exposes gaps and contradictions in the organisation's knowledge and processes, not only for the change but also for the normal operation. During normal operation, the informal organisation (Huczynski & Buchanan, 2001) will have learned to deal with these gaps and contradictions; during building and change, however, it is not possible to maintain these as hidden. This dilemma needs to be made sense of and explained within the organisation, depending on how the organisation's knowledge and processes deal with the unknown and uncertainty (Weick, 1995). In organisations that seek to rationalise such problems, these unknowns and coping mechanisms can be seen as deviations and blame apportioned or an explanation found outside the organisation, possibly within the means of building. This induces further instability in the process of building. We need, therefore, to understand what the client knows, how the client operates normally, what gaps and contradictions exist and how these are dealt with normally.

It may seem easy to collect what is known about a particular building project, but it is not: buildings are technically complex, the organisations involved are numerous and fragmented, the people involved have difficulty expressing what they mean, other people do not know what they mean, and people do not know what they do not know. All of these issues give us a larger unknown arena than construction normally acknowledges or has processes to deal with. In particular, we cannot know fully what the future building, organisation and relationships will be, but these are set against aspirations, namely hopes of what the future can be. In reality, all ends are unknown at the beginning of a project and become known only as the project transpires through the application of the means. The end is aspirational; the problem with this is that the end is not formed completely but evolves through the means of how it is being achieved. Again, the thinking of the client is key to this evolution. Ultimately, we may get not the end that was initially conceived but a better, more effective end; however, this may or may not be problematic, depending on how the client perceives its business.

The building and organisational changes are linked intimately. Each, however, can have its own drivers or disturbers. For example, with regard to the client, the business environment may change or the internal personnel may change; and with regard to the project, the ground conditions may cause delay or a subcontractor may not have enough skilled people. It is difficult to determine cause and effect in this linkage. Indeed, the initial reason for building may be a managerial strategy for an organisational change, such as downsizing or rationalisation, which interferes with the building project itself.

Change, handling unknowns, dealing with gaps and contradictions, and seeking aspirations are emotional activities. The impact of emotion is seldom acknowledged within normal business and is often suppressed rather than managed by business practices (Fineman, 2003). Emotion results from the interaction of the two dimensions of the model and it explicitly involves people; there is, however, an organisational emotion (Albrow, 1997) that influences not only people but also the operation of organisations.

The external world intervenes in all client development and in all project processes. Indeed, the distinction between the external world and the development world is rather moot, as the reason for the client development is to provide a service or to make a profit from this external world. The client environment – competitors, market, politics, regulation and finance – provides constraints for its operation and dictates partly how the client undertakes its business or service. The industry environment shares some of these aspects but also includes aspects of the industry, such as fragmentation, regulations and skills shortage. This induces a joint risk in the means-and-ends dimension.

2.4   FUNDAMENTAL AWARENESSES AND THE MODEL

The fundamental awarenesses listed in Section 2.2 are what is observed. These awarenesses involve the whole model. The model is the analytical explanation, which directs us to the theory behind the actual problems. Practitioners will align with the awarenesses and academics will align with the model, but they are in fact congruent.

The difference in perception between the client and the industry is represented in the knowledge and processes dimension. The industry sees projects as the norm, whereas clients see projects as the exception (Boddy & Buchanan, 1992). This, of course, is carried over into the means-and-ends dimension, in that the client experiences a changing situation of building and organisational development involving means, which may exhilarate or terrify them, to deliver ends, with which they may be happy or disappointed. It is almost impossible for the industry to appreciate the level of emotion experienced by the client, both during building and as the client perceives the building at the end. At times, the client may be uninterested and it is the industry's responsibility to produce a building; at other times, the client may be upset that what is happening is not within its control or what it wants. The response can be determined from the client's knowledge and processes.

The concept of building being about change in the client is within the means-and-ends dimension set against the knowledge and processes. The dissonance between these creates an issue for the client: how do the client's knowledge and processes cope with the means and ends? How does the complexity of the building situation become rationalised within the client? How does the change in the client relate to the change of building? That building is part of change in the client exposes a gap between the industry and the client and is part of the industry's and the client's different ways of seeing building. Clients seek to be in control, but they are not in control of themselves, within their organisational change or in the building. This induces emotion, which is partly in the unknown.

As mentioned, the client's knowledge and processes are not perfect: the client does not know as much about its world as it thinks it does, and the client's processes neither match completely nor coordinate what is needed. Clients, however, have difficulty in acknowledging this fact, as doing so would upset the integrity of the organisation; thus, it becomes an undiscussable issue (Argyris, 1990), which adds to the unknowns. The change, therefore, exposes the imperfections both in the knowledge (or lack of it) and in the processes, and the client may not have the means to deal with aspects outside the known.

In particular, the change exposes structural differences within the client organisation, which is not unitary in the first place. In complex organisations, different units have different objectives from the whole, and the whole is accomplished by differentiated and specialised sections. This enables the organisation to comprehend and manage its activities in units and to make it efficient. However, this gives the organisation the problem of bringing together the units into a whole. In addition, there are differences in time horizons as a result of the different objectives in the organisation's sections. Time is experienced differently in the different sections of the organisation, depending on their organisational purpose within the business environment (Clark, 2000), and this partly determines the different sections' position on the project. This causes internal tension, for example between strategic long-term views and immediate actions. Thus, ultimately, there is a conflict of purpose and a problem of coordination that requires a similar conception of time.

In addition, there are conflicts with individuals who have career aspirations. Career prospects cause differentiation between individuals. Again, as building is part of change, its delivery is part of the opportunity exploited for career progression or protection by some individuals. This opportunism induces a focus on individual needs rather than organisational or building needs. It also causes a hiding of intentions and possibly even a motivation for surreptitious action.

People and organisations have different perceptions of how ends are produced and how the future transpires. This depends on the knowledge and processes norms within the client as well as individual predilections. Many believe it is possible with dint of effort to comprehend and control the future. Others see this problem as being a fundamental metaphysical realisation of incompleteness. In the former, problems are seen as the result of mistakes or tardiness. In the latter, problems are seen as the result of evolving circumstances, thus partially outside the possibilities of rational handling. It is evident that both organisational change and building involve outcomes that are not known fully at the beginning. The majority of clients have knowledge and processes set within the former control view with a desire for stability, such that any deviance is abhorred and the system is required to return to stability. Thus, those unknowns, which come out of change, including aspirations and desires, work against this desire for stability and so induce emotions. The way in which people handle emotions can promote non-rational behaviour, and this creates contradictions with the set processes of the organisation. Contradictions cannot be acknowledged, as they undermine the integrity of organisations and their surfacing may also constrain the career progression of some personnel. We believe that a significant part of organisational change and building is not formed by initial thought but is generated through thinking processes and action (the means).

Organisational change, building and unformed futures induce a considerable degree of uncertainty, and the result is part of the unformed future. This is emotional for the client, as its aspirations may be excited at the beginning and let down at the end. Emotion is illegitimate in professional and business processes and so is often suppressed but can emerge in a radical maladaptive way if the constraints are too high and the rewards too low. This emotion is not experienced in the same way by the industry, as the industry is used to change and uncertainty because of its project environment (Boddy & Buchanan, 1992). The client, however, is not so used to this high level of change and so experiences unfamiliar anxiety and other emotions as a result of the ensuing pressure, this pressure coming from both the client's organisational change and the opportunities and threats that building provides.

2.5   MODEL FOR SECTOR ANALYSIS

In order to understand a client, an appreciation of each of the three achievements is required – the two dimensions and the external world, plus the way in which they interact and create consequences. It is evid-ent from the previous section that the dimensions are interconnected. This gives us a difficulty, as the interdependence keeps complicating the understanding. In order to explain the detail of the model, we will break down the explanation through the two dimensions, starting with the dimension of knowledge and process in Chapter 3 and then the means-and-ends dimension in Chapter 4. Some readers may want to only dip into these chapters, and so we provide an outline below, which allows us to explain how these ideas can be used for analysing a client sector. We have analysed six client sectors in Chapters 5–10.

Chapter 3 helps us understand how the client thinks normally when not building, which we refer to as 'the client at rest'. This includes the reason that the client sees the world differently than the industry. The client has unknowns even under these conditions, including its external world and its problems induced by differentiation. We use different theories of differentiation for the public and private sectors. In this regard, domain theory (Kouzes & Mico, 1970) sees public organisations with political, managerial and professional/operational domains, which have different purposes. Similarly, business system hierarchy theory (e.g. Mintzberg, 1983) sees private companies with strategic, tactical and operational strata. The problem for understanding is that these differentiated parts are meant to operate together; however, any issue or event will be seen differently through each domain or strata. At the same time, individuals have aspirations and expectations as part of their lives. Where these personal aspirations and expectations conflict with those of the organisation, they are most likely to remain unknown to the organisation, and this causes the individuals to see the world differently from the organisation.

Chapter 4 involves understanding how the client experiences change. This includes the client's basic value system, which sets what it wants (ends), and the client's basic beliefs about what needs to happen (means). The known and unknown aspects of these give us two routes of this experience: one known, rational and formal, and one unknown, non-rational and informal. These two routes give per-ceptions of the problems arising from change that are different, and this gives us a difficulty in helping the client to experience change. The client's engagement with the industry then flips between a desire for power and control and a desire for dependency.

This understanding allows us to undertake analyses of client sectors and gives us a structure for Chapters 5–10. The method of doing this is described in more detail in Chapter 11. It involves an enquiry through published and online resources and through interviewing to create an in-depth understanding of a sector and client. The enquiry involves studying three areas: the external world, the constitution of the organisation, and the sphere of influence of the building project. These have known and formal areas; however, unknown and informal areas with gaps and contradictions in operation also come out. The sector chapters are structured thus:

- ❏ *Introduction: the constitution of the sector:* this relates the fundamental aspects of the sector – the history of the sector, the sector's concerns, the structure of the sector, the major players, and some statistics about the sector.
- ❏ *The business environment: strategy or political domain:* this focuses on the detailed relationship between the sector and the business environment. A map of the client sector environment is provided as a useful representation of structure and stakeholders, so that the complexity and interdependencies can be appreciated and the effect of these on decision-making and the end experience established. The purpose of, and pressures on, the client organisation will be explored, as this determines its values and its strategy.
- ❏ *Business structure and processes: tactical plan or managerial domain:* this considers how the client's business is structured and how it sets about undertaking this. In the public sector, it is where management balances between the needs of the political and operational domains. In the private sector, it is how the strategic plan is disseminated through the organisation.
- ❏ *Business operations: operational or professional domain:* this considers how the client's business provides its products or services and the problems associated with this. This starts to reveal the unknowns in the organisation. This domain also includes how the strategy or political policies may not be deliverable or the negative pressure that the strategy or political policies may bring to bear on operations.
- ❏ *Experience of change: from unknowns and contradictions to means and ends:* this considers how the client's business will react to change, in particular how the unknowns and contradictions determine the client's means and ends during building. Finally, a means-and-ends diagram is presented, which lists the challenges facing both the client and the industry that supports the client in the organisational changes that occur during building.

❏ *Key issues:* this summarises a set of points about the sector client's business needs, processes and unknowns.
❏ *Resources:* this lists references in support of the chapter and about the particular sector and provides websites from which further information may be obtained.

2.6   SUMMARY

A model for understanding clients has been introduced. Clients and especially client sectors are extremely complicated and difficult to encapsulate. The model, however, provides a structure that enables information about clients to be sought and this information to be connected together. Although there is limited information, it is believed that this information is what is required in order to understand how a client sector thinks about building. The model is novel and on face value may seem complex and excessively abstract. However, Chapters 5–10 have practical underpinnings and demonstrate the use of the model. These six sector chapters are sufficient to enhance an understanding of and familiarity with the model.

REFERENCES

Albrow, M. (1997) *Do Organizations Have Feelings?* London: Routledge.
Argyris, C. (1990) *Overcoming Organizational Defences: Facilitating Organizational Learning.* Boston, MA: Allyn and Bacon.
Boddy, D. and Buchanan, D. (1992) *Take the Lead: Interpersonal Skills for Project Managers.* New York: Prentice Hall.
Boyd, D. and Wild, A. (1994) Action research and the engagement of construction education and practice. Presented at the 10th Annual Conference of ARCOM, Loughborough, 14–16 September 1994.
Clark, P. (2000) *Organisations in Action: Competition Between Contexts.* London: Routledge.
Dawson, R. (2000) *Developing Knowledge-based Client Relationships.* Boston, MA: Butterworth-Heinemann.
Drucker, P. (1998) The coming of the new organisation. In *Harvard Business Review on Knowledge Management.* Boston, MA: Harvard Business School Press.
Drucker, P. (2002) *Management Challenges for the 21st Century.* Oxford: Butterworth-Heinemann.
Fineman, S. (2003) *Understanding Emotions at Work.* London: Sage.

Flood, R.L. and Carson, E.R. (1993) *Dealing with Complexity: Introduction to the Theory and Application of Systems Science*. New York: Plenum Press.

French, R. and Vince, R. (1999) *Group Relations, Management and Organization*. New York: Oxford University Press.

Goldratt, E.M., Schragenheim, E. and Ptak, C.A. (2000) *Necessary But Not Sufficient: A Theory of Constraints Business Novel*. Great Barrington, MA: North River Press.

Heirs, B. and Pehrson, G. (1982) *The Mind of the Organization*. New York: Harper & Row.

Huczynski, A. and Buchanan, D. (2001) *Organizational Behaviour: An Introductory Text*, 4th edn. Harlow: Prentice Hall.

Kouzes, J.M. and Mico, P.R. (1970) Domain theory: an introduction to organisational behavior in human service organisations. *Journal of Applied Behavioural Science*, **15**(4), 449–469.

Mintzberg, H. (1983) *Structure in Fives: Designing Effective Organizations*. Englewood Cliffs, NJ: Prentice Hall.

Sherwood, D. (2002) *Seeing the Forest for the Trees: A Manager's Guide to Applying Systems Thinking*. London: Nicholas Brealey.

Weick, K. (1995) *Sensemaking in Organisations*. London: Sage.

3 The Client at Rest

3.1 CLIENT'S KNOWLEDGE AND PROCESSES

This chapter sets out our model of the client before they build. We will refer to this as the client organisation 'at rest'. This is summarised in Figure 3.1, which is part of our overall model introduced in Chapter 2. We will consider the normal organisational structures, behaviours and problems, some of which are unknown, the fact that organisations are composed of people, which also creates a set of unknowns, and also the route to coping with the unknowns.

In the model, we consider the different ways in which client organisations act in the world, determined by their knowledge and processes. Knowledge is the information that the client has at its command concerning its business environment, strategic direction and internal structure. Processes are the formal systems engaged in by the organisation, both in procedures and decision-making. This is adopting an information-processing (Drucker, 2002; Winch, 2002) and knowledge-management (Sveiby, 1997) perspective of the world, although with the aim not of managing knowledge but of understanding the client's actions in terms of knowledge. Subsequently, it may well be a strategy to undertake business and to deliver client satisfaction by managing client knowledge (Dawson, 2000), and this will be dealt with in the final chapter. The way in which a client thinks about the world is fundamental to the way that the client sees (and experiences) the world and makes decisions within it (Heirs & Pehrson, 1982). Thus, it is the starting point of understanding the client's engagement with construction, how the client works with the industry, and how the client experiences the end point, giving the client its priorities and values

Figure 3.1   Model of organisation at rest.

that contain what is important and what is good and bad and relating to what will satisfy the client. In a contingency model, which we are using, the client's way of thinking is formed by external conditions (or, at least, it must be sustainable in external conditions), its history and its particular operational needs (Emery & Trist, 1969). This is exhibited in both the organisational structure and behaviour.

We have made this chapter relate to a number of areas that are needed in Chapter 4, where we will consider the effect of building. Key to any organisation in terms of both its operation and its identity is its purpose. We will first describe the key idea that a client's purpose is not building and so its world is different from that of the industry; therefore, the client thinks differently, sees things differently and even appreciates buildings differently from the industry. These differences mean that we need to understand organisations in general and what makes them different. This relates to what makes organisations viable within their external environment. We will describe models of organisational structure and process that help us to understand client organisations, making a common distinction between public-sector and private-sector clients, as these relate to their external worlds differently.

In preparation for Chapter 4, we are also interested here in aspects of client organisations that will be put under stress when they come to build and that we identify as organisational unknowns. This is only a definition and, as we will explain, these may not be unknown but only hidden. Importantly, however, these unknowns cannot be part of the rational organisation; they create what we call a rationality gap, i.e. the gap between what an organisations pretends about its operation and what really happens. Therefore, we cannot make the normal assumption that the client's business works perfectly and has all the information available in order to make decisions. The significance of these unknowns is seldom acknowledged and seldom considered by clients or the industry during construction; only the effect of the unknowns is managed. These unknowns are part of normal organisational operations. There is no perfect organisational structure that operates perfectly: all organisations have knowledge gaps and contradictions. For example, a client may not know about its business environment because of its vastness and because it is changing. A client may not know about its internal systems because they are not ideal and because people hide problems. A client may not know whether its decisions are right. All organisations experience these problems and develop ways of coping in their normal existence. The desire to be in control of these unknowns and to return to stability is currently encapsulated in risk management. Thus, as well as considering what a client knows and how it procedurised its thinking in order to make effective action, we must also consider what the client does not know and how its processes cope with this or, indeed, create difficulties. In a strange way, unknowns create more unknowns, depending on the way in which an organisation responds to unknowns. These aspects will be dealt with alongside an awareness of organisational culture. We use this concept pragmatically in order to discuss differences in organisations.

Finally, we need to see organisations as being composed of people. Indeed, when organisations build, this allows individuals' aspirations to come to the fore. So, in the same way that an organisation is under stress, so are its people. From an organisational perspective, people can be regarded as one of the unknowns, because they have alternative agendas, such as their careers, and because they have emotions.

communications is more sophisticated and accommodates the ideas of perception but also introduces another element of the social psychology of communications, namely expectancy. Expectancy occurs as the first element in the processing of the message, whereby the individual receiver seeks out particular stimuli from the environment. The motives and desires of the individual are influences at this stage in the process, e.g. if the client's expectation of architects is that they are profligate in spending money, then the client will be sensitive to and suspicious of any indication of additional expenditure induced by the architect, even if it is a good suggestion. Britt (1979) refers to the effect of expectations and how it causes people to act towards stimuli:

> The individual has developed certain attitudes about the stimulus and if the stimulus is in line with his expectations, then a higher relevance value will result. If it is quite contradictory to his previous attitude, a stimulus could arouse a great interest, incongruity, or be rebuked and cast aside, in other words not perceived.

This allowed Boyd and Kerr (1998) to prepare a model of communications (or miscommunications) between client and consultant across the perceptual gap, as shown in Figure 3.3. There is often an assumption

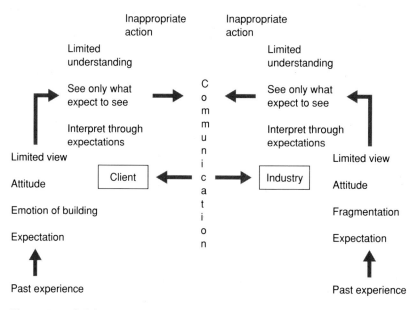

Figure 3.3   (Mis)communications across the perceptual gap. Adapted from Boyd and Kerr (1998).

that what is being said and what is being heard are the same. This is unlikely to be the case, as the message goes through complex stages, which change its meaning and value, before receipt by the receiver. Clients react not to the immediate situation that confronts them but to their perception of the situation and their mental image drawn from their experience of the world. Client expectations cause them to react in a certain way, which may be wholly different from the expectations of the industry. The message is, therefore, filtered through the client's beliefs and assumptions.

There is a strong link between expectations and attitudes. Hay (1993) illustrates this link when looking at the interrelationships at work using a transactional analysis approach. Hay has developed an attitude labelling model, which shows that the forming of attitudes is based very much on expectations and that a reinforcing loop exists, which feeds the perception of 'their behaviour' based on the restricted samples that are observed. The industry's behaviour is always viewed from the perspective of the client's restricted analysis. Although the viewer knows that they do not witness the whole of the other's behaviour, conclusions are drawn about 'their' attitude to the world. Once the observer has determined in their mind what the other's attitude is, then they will predict the other's behaviour through this expectation. Armed with this, the viewer edits out things that do not fit in with their expectations and reinforces the loop by selectively identifying further restricted samples of behaviour that do fit in with the viewer's expectations. Thus, expectancy and attitude can build up in a relationship to prevent successful communications. Experienced clients may have developed expectations and behaviours that are negative to the progress of a relationship with the industry. Such an attitude may generate the behaviours they fear most because these are the only responses that are normally available. Thus, change requires changing expectations and exceptional behaviour in situations and can be likened to turning the other cheek when attacked.

Client's view of buildings

It may come as a surprise to the construction industry to know that clients do not think much about their buildings most of the time (McGregor & Then, 1999). If a client needs buildings regularly as part of its business, then it will probably have a property department. This division will certainly think a lot about buildings, but it is not representative of the client. Buildings to most business directors and managers are almost a nuisance; at best, they are a necessary evil. They think about the buildings when they are a problem, either because they are not functioning and people are complaining or they are costing too much money. Buildings fit very much into the motivation category of hygiene factor (Herzberg, 1966) in that organisations will strive to achieve their needs through buildings and will notice when their needs are not met; if the problem is solved, however, they lose interest rapidly. When a building is operating well, it is not noticed explicitly. Even the aesthetics of buildings can illicit such a response and may be maintained in an organisation's awareness only if it is providing publicity and identity. Few business textbooks mention buildings other than as an asset or liability, and thus few managers are taught to appreciate the significance of the quality of their built environment for their organisational productivity (McGregor & Then, 1999).

Buildings are also seen as a problem by clients because they have a history of being problematic, both in their operation and in their construction. This induces a degree of trepidation in business managers who have to do something with buildings. Specific organisations may have particular stories to tell about a past problem, which heightens the fear, even if this was an isolated incident. Thus, buildings are problems for the organisation and, moreover, cost money without the organisation appreciating what the building is doing for the organisation.

3.3   ORGANISATIONAL KNOWNS

In order to understand clients, we need an understanding of organisations in general. Organisations are patterns of association that have an identity that can be recognised (Weick, 1995). In most cases this is easy but, as we will see later, confusion can arise from the differentiated nature of organisations. Organisations also have an existence in time (past, present and future) over which this identity has to be maintained; that is, they have a history to be told. This allows us to see some of the concepts with which we must work in order to understand organisations – patterns of association, which include structure, roles, procedures and relationships; identity, which includes purpose, product, service, people, procedures, and icons such as buildings and logos; and time, which includes history, performance (efficiency), relationships and strategy. Regarding organisations in these very abstract terms allows us to perceive any organisation; however, we must also be able to see the particular, as this is what we work with.

There is an extensive literature on organisations, which falls into a number of theoretical camps (Huczynski & Buchanan, 2001), each adopting different viewpoints on how the organisational world works. We use a system perspective (Flood & Carson, 1993) based on constraints both externally to the situation as in contingency theory (Huczynski & Buchanan, 2001) and from the constitution of the system (Goldratt *et al.*, 2000). This allows us to both look inside organisations and to see outside them; but most importantly, it allows us to see organisations as an integrated whole. This is also a common theoretical perspective for construction (Barrett & Stanley, 1999; Walker, 2002), which aids us when we come to look at them together in the next chapter. It was also the perspective used in the first attempts at understanding clients (Crichton, 1966). Finally, a system perspective allows us to see problems in organisations and potentially to offer some suggestions of how these can be overcome. The system perspective is fundamentally about seeing organisations as wholes within environments; however, there are differences in beliefs about how we know about the world (Checkland, 1981). Although fundamentally we are adopting a soft systems approach (Checkland, 1981; Flood & Carson, 1993), whereby knowledge is socially constructed (Easterby-Smith *et al.*, 1994), we will use pragmatically systems engineering ideas to assist our explanation.

The work on viable systems gives us a tool to describe and analyse the normal operation of organisations as they seek to have an efficient operation but also to have a sustainable future by matching the developing business to their environmental. The viable system model (Beer, 1985; Espejo & Harnden, 1989) offers a way of understanding both functional decentralisation and cohesion of the whole. It is based on cybernetics, which concerns communication and control in complex organisations. A viable system needs to have five subsystems in place in order to operate effectively in its business environment. These subsystems, shown in Figure 3.4, are:

❑ *Implementation:* produces the organisation's products and services and adds value into the business environment. There may be a number of subsystems (DA, DB, etc.), which deal with different aspects of the business environment (EA, EB, etc.).
❑ *Coordination:* acts as the interfaces between an organisation's value-adding functions and the operations of its primary subunits.

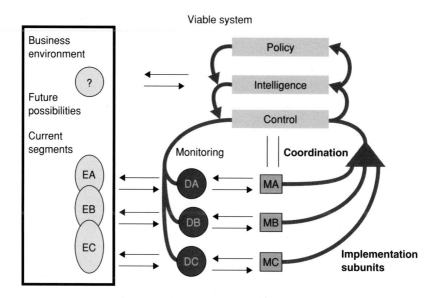

Figure 3.4 Viable system model.
Adapted from R Espejo and R Harnden *The Viable System Model: Interpretations and Applications of Stafford Beer's VSM*. Copyright 1989. John Wiley & Sons Limited. Reproduced with permission.

❑ *Control:* the intermediary between policy and implementation, which thus amplifies instructions and shows accountability to higher-level management while negotiating for resources.
❑ *Intelligence:* provides information on the market and other factors that could affect the future, while also selling the organisation into its environment.
❑ *Policy:* acts as an arbitrator between the intelligence and control functions and from these provides a clear overall direction and purpose to the whole organisation.

The model also uses the concept of recursiveness to describe the way in which each of the nested subsystems needs to have similar self-organising and self-regulatory characteristics in order to function.

The external world

As shown in Figure 3.1, the external world contains the organisation. We are dealing with an open system (Flood & Carson, 1993); thus, the degree of this influence can be quite extreme, defining whether an organisation strategy is effective and whether operations are efficient, and ultimately causing the organisation to succeed or fail. The viable systems model shown in Figure 3.4 makes us aware that an organisation works only with small parts of its environment and thus has knowledge only of this; this knowledge may be held in different part of the organisation. We will now consider some of the important aspects of that environment. The acronym PESTLE, standing for political, economic, social, technological, legal and environmental, is often used to analyse the external world (Huczynski & Buchanan, 2001). *Political* refers to local and national government action, including both regulation and policy direction. *Economic* refers to the wider financial context, such as interest rates and taxation. *Social* includes the way in which society is differentiated, and how the different groups behave in both a market and a human resource sense. *Technological* refers to the impact on or substitution of products, as well as production and delivery aspects. In many studies, the technological base defined an organisational structure (Huczynski & Buchanan, 2001). *Legal* sets the constraints on the legality of the operations within an organisation, and this can have an impact on a number of different areas within the organisation, from finance to health and safety. *Environmental* refers to the conditions set nationally and locally concerning pollution and resource use. This is not only a legislative area but also a public relations and moral area.

The external world also has other organisations that affect the client (Porter, 1985). These involve the market, where people buy products or use the client's service; the competition, which could take away from the organisation the current and future market; suppliers, who are part of a value chain and thus dependent on the client but are also in competition with the client for higher prices; and finance from which the capital for development and continued operations comes. Together, these forces are acting on an organisation at all times. Indeed, in a contingency theory model, it is what makes an organisation successful or not. However, the factors are not fixed, and so the client organisation has to be aware of how the external world is changing.

Purpose and strategy

A successful organisation provides a relevant function within its business environment using efficient and coordinated internal processes while at the same time learning how it needs to change in order to continue to do this or to do this better in the future. Organisational purpose then has to be aligned to the environment. This is set within the policy function in the viable systems model; however, this assumes a rational model, and there may be other determinants of what really happens. In our attempt to generalise about organisations, we run into the differences between public service organisations and for-profit businesses.

One difference between private and public organisations is the source of the purposive activity. Clearly, both have an overarching purpose almost set by their environment, private (e.g. multinational oil company, national food retailer, local developer, small engineering company) being the pursuit of profit and public (e.g. local authority, government department, health trust, school) being the pursuit of social operation and support. They are also distinguished, however, by the way in which each organisation works to create this purpose and how they operationalise this. They both have processes of decision-making and of making strategic choices. In the private sector, this is very much set internally. Although this might be on a hierarchical basis and thus distant from the point of action, it is still an internal decision within a broad context of the law and the working of the market. The profit centre can say what it wants to do (business plan) and the hierarchy agrees it. However, in the public sector, strategic decision-making is derived more externally. The political context sets the agenda and almost declares a detailed action that has to be implemented by the organisation. The organisation's discretion is limited to an inter-pretation of time and presence.

Both public- and private-sector organisations will have a strategy, namely the long-term aims that fulfil the organisational purpose. The list of strategic aims presented in Table 3.1 provides examples of these. Again, these will be generated internally in private companies but constituted politically within public-sector organisations. The strategic aims will be utilised in a strategic plan, which will present ways in which the aims are to be achieved. The elements of a strategic plan are presented in Table 3.2. There are a number of strategic analysis models, such as Porter's (1985) five forces, which relates together these aspects in order to assist in making decisions, although most of these models are concerned with private organisations.

Organisational differentiation

All large organisations differentiate themselves internally into layers and departments with different purposes, and these develop different views

Table 3.1   Strategic aims in public and private sector organisations.

Private sector	Public sector
Profit level	Accountability
Improved financial ratios	Transparency
Efficiency	Social involvement
Productivity	Democratic decision
Market position/share	Equity
Increased dividend	Fairness
Customer retention	Employee welfare
Customer satisfaction	Reduced complaints
Corporate governance	Effective service
Corporate social responsibility: employee welfare/environment/community	Reduced staffing levels
	Benchmark targets
Value for money	

Table 3.2   Elements of a strategic plan.

Strategic aims	
Market: definition of market	Statistics about market
	Explanation of market: trends, cycles, business drivers, customers' priorities and needs
	PESTLE analysis
Business structure and operations	Existing customer
	Products and services: production and delivery
	Market methods
	Sales methods
	SWOT analysis
Competitor analysis	
Actions to achieve aims	
Resource implications	
Budget and finance	

SWOT, strengths, weaknesses, opportunities, threats.

on what the organisation does and have different objectives (Huczynski & Buchanan, 2001). This is useful for both the organisation's purpose and the organisation's performance, as it allows each part to do its job better and with fewer resources. However, it gives the organisation a problem in that these layers and departments have to be integrated, both for the wider task of the organisation and for its identity. This activity is referred to as coordination in the viable systems model and is often a differentiated function in its own right. This differentiation also gives the organisation another problem in that external organisations, e.g. suppliers and customers, see different parts of the company as having different objectives, which can be contradictory.

The most common vertical distinction in a formal organisation would be strategic, tactical and operational management (Huczynski & Buchanan, 2001), as shown in Figure 3.5. Mintzberg (1983) develops this more fully with five areas, but this overcomplicates our purpose and we will stick to three levels. We will call this Business Systems Hierarchy Theory to distinguish it from Jacques' (1989) Stratified Systems Theory. Each of these strata has a different time horizon for thinking about the organisation (Jacques, 1989) and different objectives. Strategic refers to longer-term thinking (over three years) and is concerned with developing a stronger share value by selecting the different directions that an organisation could take within its market. Tactical refers to medium-term thinking (one to three years), where the overall structure and processes of the organisation are determined in relation to the strategy and in relation to the operational capability delivering profit and productivity through the efficient use of people, resources and suppliers. Operational refers to short-term thinking (up to a year, but can be daily), where problems of production or delivery are resolved, and immediate improvements made to efficiency, in order to deliver a good or a service to the client's customer. Clearly, these relate

Figure 3.5 Vertical differentiation in organisations.

to the viable system; they are also, however, a justification for the organisational hierarchy that differentiates the organisation on importance
and reward. Organisations have become flatter and less structured since
Jacques (1989) introduced this conception, and it may well be that any
one of the client's managers is undertaking all three levels of concern.

The public sector

The pressures on public-sector organisations differ from those on private
organisations, and this means that a different model needs to be used.
The public sector, within a mixed economy, serves both as a demon-
strator of a democratic social system and as the deliverer of shared
and socially supportive services. Domain theory (Kouzes & Mico, 1970)
allows us to rationalise the divisions and decision-making within public-
service or pseudo-public-service organisations. It sees three domains
of operation – political, managerial and professional – as representing
the constitution of the public sector, as shown in Figure 3.6. These have
different inner logics, working patterns and values (Talbot, 2003),
which work in a degree of tension. It is this vertical tension that is not
so pronounced in private organisations. The concerns of the political
domain are what services will be provided, the priority of these and
the resources that these services attract. The method of operation has
to address the wide range of stakeholders involved in any service,
and the decision-making can involve lobbying, negotiation, power-
brokering and selling. In the resource-limited world of the public
sector, this domain also decides on what services will not be provided,
with the consequences often being passed to the other domains. The
managerial domain is a bit like the tactical level, in that it is concerned
with implementing the decisions of the political domain. The method
of operation here tends to be bureaucratic, based on producing rational
procedural documents that match the political purpose, involve accountable and calculative decision-making to demonstrate the public service
value, and be hierarchical as regards its operations. The professional
domain is about operationalising the managerial direction but is within
norms set nationally by professional institutions. The way of working
tends to be about implementing rational procedures, involving a high
degree of correct personal interpretation, which is used to define the
quality and timeframe of the service provided.

This distinction between domains is changing. The direction of public
activity is being moved to a situation more like a private-sector model,

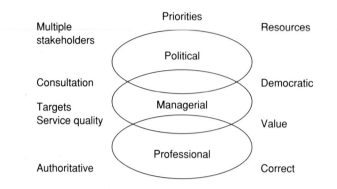

Figure 3.6   Domain theory: vertical differentiation in public organisations.

with higher-level aims, with ideas coming from individuals locally, and with good ideas supported by the hierarchy. These changes emerged during the last quarter of the twentieth century (Ranson & Stewart, 1994). Before the 1970s, the public sector was based on a professional bureaucracy whose role was administrator of policy. This notion of public administration involved experts providing for a social democracy for which the public consented. This was problematic, because people were not consulted and also because of its high cost. Thus, in the 1970s, there was a move to public corporatism. This involved a more efficient delivery of service against more structured objectives set higher in the public corporation and not left to the professional administrators. Again, the system operated through highly rational processes, but it also concentrated power within the managerial domain. The current move is to neo-liberalism, where an explicitly private-sector model of organisations has been adopted. In this, public services are delivered to customers, who have rights in relation to defining the service and its quality, but at the same time the services operate within strict budgets for the delivery of this service set by the political domain. There is a focus on value and efficiency and a further derogation of the professional domain in contrast to its status in the earlier public-administration model. Other ways of providing these services are now being explored, such as private finance initiative (PFI) and regulated private monopolies, in order to seek more efficiency. There is also a new agenda for freedom of choice, represented by public-choice theory (Hughes, 1998), which explores a regulated market delivery of public services based on the achievements of targets.

Processes and control

One of the key issues about organisations, whether public or private, is that they seek to control tightly and optimise their operations. These operations are what organisations undertake every day, and this familiarity allows the organisation to have a high knowledge of them and so create rational models of operations that align well with reality and can be used for control. Such models are often devised through system engineering (Checkland, 1981), which sees organisational processes as a series of flows with appropriate feedback allowing for immediate control. These models have their roots in Taylorism and see the world as a machine (Morgan, 1997) that is compliant to instruction and rules. Operational managers spend much time optimising these flows and the rules for action. They then seek to manage the operational situation by ensuring that the operations are stabilised at this optimum and deviations are drawn back to this optimum. This develops the natural way of thinking within the organisation to be concerned with ensuring a return to stability at any evidence of deviation. The organisation feels safe when it is working in this automatic way, and this can be monitored easily by regular reporting of performance using suitable metrics set within the task flow. The organisational purpose then becomes the management of these metrics in order to maintain the internally stable system. People are recruited, trained and evaluated for this stability, which becomes part of the organisational culture.

Organisations differ in the degree to which they have naturalised these stability-seeking processes. Such stable systems may become not only a part of operational management but also part of both tactical and strategic management. Organisations may find it difficult to deal with changes in their environment as they are constituted internally in order to ensure stability. This also aligns with attitudes to risk, as this relates to events that occur outside the operation of the stable system. The very definition of risk in a stability-seeking organisation tends to be as a negative occurrence that needs to be avoided or managed out, as it delivers outcomes that are not planned. We take these areas of misalignment between an organisation and its environment and inside an organisation as organisational unknowns, which we introduce in the next section. We also consider a concept of uncertainty, which is often defined as a further degree of unknown because it has not been preconceived, let alone mitigating actions put in place.

3.4 NORMAL ORGANISATIONAL UNKNOWNS

Organisations from the outside look competent, coherent and in control; that is, they work well with the right resources, their divisions work together and understand how they fit into a whole, and they know where they are going and how to decide on this. This is not really the case, however, and the world of an organisation is unknown for a number of reasons; this is represented to the right of our model in Figure 3.1. Much has been written about what organisations know, but very little has been written about organisational unknowns. This is a philosophical topic and as such is open for debate and differences of opinion. The unknown is acutely important in organisational change and in building. We introduce it in this section as being part of normal organisational operation, but we return to it in Chapter 4 when we consider what happens when an organisation builds. Many notions of the unknown involve the idea that it is possible to know and also, in that sense of discovery, that if we looked we could know. This may not be the case, as some knowledge may not be knowable (Atkinson & Claxton, 2000) and some may not be in existence; however hard we try to look for it, it may not be there yet (Stacey, 2001). In this book, we make a distinction between hidden unknown, latent unknown and unformed unknown. The first two encompass the idea of the possibility of discovery; the latter, however, involves a different idea, namely that things come into existence that were not there before (Stacey, 2001). Texts on knowledge management are concerned mainly with using better the knowledge that is there; however, there are aspects of knowledge creation and different types of knowledge (tacit and explicit) that suggest that simply managing known knowledge may be too simple an idea (Nonaka and Takeuchi, 1995). A list of normal organisational unknowns is presented in Table 3.3.

In our definition, then, *hidden unknown* is knowledge that is known but not revealed. The intentionality or mistakenness of this lack of revealing involves not the knowledge itself but interactions between people. This sociology of knowledge asks why organisations or people do not convey what they know for reasons that benefit, or apparently benefit, them.

The *latent unknown* is knowledge that can be known but is not realised as being significant to the situation; that is, the connection has not been made or the question has not been asked. This can be because we have not discovered it yet. It is a connection between the situation

Table 3.3   Organisational unknowns: gaps, contradictions and uncertainties.

Strategic ambiguity	Why are we doing this? May be hidden, as there may be agendas that are not declared; may be latent, as the significance of the decision has not been worked through
Discontinuity with tactics and operations	This is hidden in the sense that problems are appreciated (or anticipated – 'Look what happened last time') but, because of selling and the need for being positive about plans, the problems are not discussed. This can be latent, in that the significance of the change has not been thought through; it could be avoided or not considered important or known what to do. It could be unformed in the sense that we do not know how it will work until we try to make it work
Divisional communications	Inadequacy of information or of interpretation
Internal conflict	Opposing objectives or values
Uncoordination	Activities that mismatch or reciprocal interdependence (Thompson, 1967)
Assumptions	Basis of making decisions that look like previous decisions
Internal cycle mismatch	Efforts and objectives change, depending on time in the organisation's year or week or day ('a Friday-afternoon job')
Careers	Objectives of individuals against other individuals or against organisation are in competition
Illness	Affects communications and storage of information as well as emotional state, as the responsibility or duty is not there – thus creates unformed events
Personnel changes	Like illness, but with knock-on organisational implications and even inducing others to think of their positions as unintended consequences
Selling of change	Promotion/propaganda tends to overstate possible future. This creates a hidden awareness of impossibility coupled with a potential reaction against being patronised. The consequences are unformed, depending on circumstances

Table 3.3 (*Continued*)

Organisational emotion (climate)	There is a group feeling that non-linearly can create panic or euphoria. This may be latent, but also its propagation is unformed
External change	The goalpost moves – 'Who stole my cheese?' It is believed that this is latent in the external system within PESTLE. If we looked in the right way, we could see it coming
Misalignment with external	This is latent and unformed. If it is a new product, then the alignment to the external is not known or formed. If it is an old product, then it is latent in the sense that there must be evidence of demise, but this is not read as such. (It could be hidden in the sense that someone knew but was not saying or not being heard.) It could also be external change
Financial conditions	Although covered by external factors, this has been stated separately because it is so important in organisation stability and also because it is an interaction between external states and internal state, e.g. cash flow may be the problem rather than overall profitability, which might be due to others experiencing difficulties either from external change or from internal opposing objectives
External time cycles	Like internal cycles, where there is a mismatch between what happens inside and what happens outside
Discovered opportunities	If physical, then these were clearly latent; if organisational, then they were unformed and created from the flow of the means. Time allows opportunities to be developed but also for things to go wrong
Events	Things that happen appear latent but are most often unformed and induced by complex conditions of external and internal systems

PESTLE, political, economic, social, technological, legal, environmental.

and knowledge. Somehow, the situation is different from our knowledge. Once found, this knowledge is there and in most cases can be discovered through investigation. The notion of unknowable knowledge such as tacit knowledge and skills is a typical feature that cannot be communicated or explained (Atkinson & Claxton, 2000) although its outcomes can be described.

Finally, the *unformed unknown* is knowledge that is not in existence yet and will become known only through the sequence of events that led up to its becoming known. This is the most complex of knowledge and intimately involves the circumstances surrounding how it comes into being and the time over which this takes place (Clark, 2000; Stacey, 2001).

Rationality gap

The classical theories of business involve a large amount of rational planning, which identifies what should be known about organisational achievement. This comes from a scientific view of the world that was drawn into economics and administrative management (Stacey, 2001). As this is not totally realistic in most circumstances, we are left with a gap, which we shall call the rationality gap. This gap is represented in Figure 3.1 between organisational knowledge and processes and organisational unknowns and coping mechanisms, starting with normal organisational gaps and contradictions; it is made worse, however, by the fact that it cannot be acknowledged. Some organisational processes are in place to ensure that this gap is as small as possible. For example, quality-management systems (McCabe, 2001) and business process mapping (Wilson and Harsin, 1998) are both rational techniques that organisations use to overcome some of these endemic problems. In addition, failure modes may be searched out and mitigation put in place to control their effects (McDermott *et al.*, 1996).

Most organisations struggle with this rationality gap (Quinn, 1988). The word 'rational' means that we have a logical explanation of events (Stacey, 2001). Organisations are drawn into this process of rationalising their operations in order to find reasons for the gaps and contradictions and to identify processes that are not delivering what was wanted, i.e. what is unknown. It is easier to find rational explanations when technology does not deliver our expectations than when organisations and people do not deliver; however, we often apply the same thinking to both situations, and this can be problematic, as rational explanations for organisational behaviour are often spurious (Stacey, 2001). These explanations work depending on the repeatability of the processes being studied and the compliance of the workforce to improvements. In addition, such explanations require a good relationship between the management model of the organisation and the way in which the organisation actually operates. Organisations are less easy to know about – and even if we have rational explanations, we cannot be sure that they are true. There are always mismatches, and these are coped with differently in different organisations. It is difficult to distinguish between what is genuinely unknown and what has been caused by tardiness of the means in achieving the ends of our rational plan. There is temptation in some organisations to regard it all as tardiness; indeed, such organisations assert that if their plans had been created correctly and implemented properly, then the plans would have been achieved. This response creates additional unknowns, not only because of the gap but in the explanation of it. An important aspect of this response to the rationality gap is represented by the distinction between an organisation's 'espoused theory' (what it says it does) and its 'theory-in-use' (what it actually does) (Argyris, 1993). A real gap exists between espoused theory and actual behaviour in most situations. This gap results from the way in which individuals within groups become conditioned to hide their assumptions, even when everyone knows that they are assumptions; they both choose to act as if neither of the parties knows anything. These defensive routines result from a fear of blame and are part of the emotional problem of the unknown in organisations (Argyris, 1993). We often use the word 'irrational' to describe the actions of someone who does not meet our ideas of rational and logical (Brunsson, 2000). There may be a logic, however, and it is hidden or one with which we do not agree. Such pejorative responses come out of the perceptual gap identified in Section 3.2 between clients and the industry, who see the world differently. Thus, what is described as irrational to one is rational to another. Using the word 'non-rational', i.e. not explainable or unknown, rather than 'irrational', i.e. stupid or perverse but known, allows us to see this gap in perception as an inevitable problem of difference.

Constitution of organisational unknowns

A major point, which will be taken up in Chapter 4, is that when organisations build, they become more vulnerable to their normal unknowns, whether hidden, latent or unformed. The roots of these unknowns listed in Table 3.3, whether gaps, contradictions or uncertainties, may be in strategy, in structure, in process, in operations and in people. As regards the normal operation of organisations, most unknowns are in our categories of hidden or latent unknowns. In this sense, the organisation can work on them if it can appreciate them as unknowns, so as to bring them into the known arena. Thus, working on these unknowns may be controlled by the organisational processes (e.g. continuous improvement) that seek to overcome the hidden unknowns and to uncover the latent unknowns. However, all organisations have unformed unknowns concerning their future existence, as they work through their problems and success. These unformed unknowns are caused by the external environment, from gaps and contradictions in internal structure and processes and from the social nature of organisations.

A viable organisation undertakes activities that are in demand by the environment. Unknowns exist in the environment, relating to the continuing viability of these activities. This can be because an activity is obsolete or because a competitor is doing it better. Although a strategy may be effective theoretically, it has to be delivered by the organisation and this capability in itself is uncertain. This is displayed in conflicts between improving the organisation in the short term for immediate returns, and thus forgetting about the future; or, conversely, developing the organisation in the long term, which reduces output in the short term but makes the organisation sustainable in the long term.

The problems of structure and process are inherent in all organisations, because of the necessity for a differentiated structure and from the impossibility of having processes that are completely consistent and universally applicable. The notion of differentiation within organisations was introduced as an organisational known in the last section; however, it is the problem of integration that produces many unknowns (Lawrence & Lorch, 1967). Lawrence and Lorch's ideas suggested that as environments became more uncertain, then organisations become more differentiated in order to respond flexibly to the situation. However, this creates greater goal divergence between the differentiated sections and so leads to internal conflict. We shall deal with this in more detail in the next chapter, when we suggest that organisational change involving building creates this uncertainty. Espejo and Harnden (1989), using the viable systems model, identified coordination as a problem in many organisations because it is regarded as top-down direction and control rather than as a true enabler of mutual adjustment. The coordination role has insufficient information compared with the implementation function to be effectively directing the operations from the outside. In a similar problem, Espejo

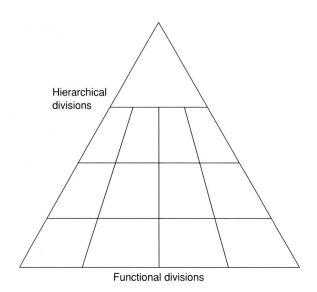

Figure 3.7   Creation of operational islands.

and Harnden (1989) see control from policy level trying to reach down too far, with even more disastrous consequences, not only because of the lack of information but also because of the unintended consequential destruction of trust in the lower levels and their abdication of their role in ensuring mutual adjustment between the intelligence and operations level.

We have also identified a differentiation by function across organisations, e.g. finance, production, marketing, etc. These tend to be given a spatial separation in floors or buildings, so when they are put together with stratified systems, organisations have what Rush (1991) refers to as operational islands, as shown in Figure 3.7. The vertical strata refer not only to a time horizon but also to a relationship with the outside world or the business environment; this model then starts to become useful for our purpose. We might also add a power dimension to this, as many organisations have separated managerial power and functional power.

Operational islands create gaps and contradictions. Gaps occur at boundaries between sections or processes and may be latent to those near the situation being interpreted as a failure in another section. What is not available is an overview of the whole process in order to establish what is latent when a new event has to be dealt with by the differentiated organisation. In addition, it is often hidden by the organisation dealing with it as an exception, thus not exposing the system failure. Contradictions occur where two or more opposing ideas are held simultaneously (Talbot, 2003). Each idea is known about, but there is a disagreement in logic between them. Often, these contradictions are hidden and not talked about or considered significant.

From the latter, it is clear that the real operating organisational structure is important in the determination of internal organisational unknowns, as it is this coordination problem that the organisational processes have to bridge. There are a number of conceptions that help us establish the way in which knowledge and processes deal with unknowns. Transaction cost analysis (Williamson, 1975) models such problems as impacted knowledge and sees two causes of this, namely complexity and opportunism. Complexity is part of our bounded rationality (Simon, 1976), which limits us individually and in a group from having knowledge of a situation, providing latent knowledge in our definition. Opportunism is the action of a group or individual seeking advantage from its knowledge. This knowledge is available if it is searched out. Thompson's (1967) work on organisations sees the organisational structure of operations in relation to task as being critical, namely that the degree of interdependence of processes, whether pooled, serial or reciprocal, determines how the information flows have to be managed in order to produce successful coordination of specialised departments.

Culture and the informal organisation

The above analysis identifies the unknowns in the formal organisation. From the outside, an organisation appears to be a rational homogeneous and goal-directed enterprise; however, all organisations also have an informal composition that is non-rational, unknown and centred around people. Many of these informal aspects can be seen as part of the culture of organisations, which are hidden to the outside and, to some extent, unknown to the inside. The informal organisation is formed by people in sentient groups being drawn together around their personal identities, affinities and preferences (Miller & Rice, 1967). Aspects of the informal organisation are presented in Table 3.4. This is sometimes referred to as the organisational iceberg, with the informal organisation being the nine-tenths of the organisation that is hidden, thus emphasising the importance of this unknown.

Table 3.4   Aspects of the informal organisation.

People-centred	Organisation-centred
Patterns of interpersonal, group and divisional relationships	Power and influence patterns
Workgroup sentiments, norms	Transmission of management style
Perceptions of linkages, trust	Emergent policy
Risk-taking behaviour	Organisational value systems
Role perceptions	Divisional intentions and aspirations
Emotional feelings, needs	
Quality of working relationships	

Adapted from Huczynski and Buchanan (2001).

These informal relations not only accommodate the needs of people but also provide a mechanism for assisting or hindering the formal operation in handling its stresses. From the inside, an organisation has complexity (not understanding itself), uncertainty (not knowing where it wants to be) and conflict (disagreement on what and how things should be done, plus individual and group career advancement). All these unknowns induce a feeling of uncertainty; the way in which an organisation handles its uncertainties, which occur during its normal operation, can be usefully seen as part of its organisational culture. Indeed, Hofstede's (1994) analysis of culture identifies uncertainty avoidance as an important dimension of operation. *Uncertainty avoidance* is the extent to which the culture feels endangered by unfamiliar happenings. The other dimensions of Hofstede's analysis provide us with a distinction between different organisations. These dimensions are: *power distance*, i.e. the extent to which power is dispersed across the organisation; *individualism*, i.e. the degree of importance of an individual's requirements compared with the group's needs as a whole; and *masculinity*, i.e. the extent to which actions show evidence of masculine competitiveness against feministic cooperativeness.

The concept of culture is complex, and many authors criticise it as being an inadequate theory for real organisations. Wright (1994) states that cultures are neither universal in a community nor static, and thus decisions occur by a series of negotiations. Morgan (1997) avoids this methodological debate by stating that he uses differences in organisations in a metaphorical sense, and we will adopt this defence as well. The simplicity of Hofstede's categories allows us to focus on important aspects of differences between organisations.

Organisations that are uncertainty-avoiding tend to eschew risks and create complex rules, in both the formal and informal organisation, in order to protect themselves from difficult situations. Organisations with low uncertainty avoidance are more comfortable with ambiguous situations; they are also more relaxed about change and innovation. Organisations with a large power distance have much greater stress on their hierarchies, such that informal political manoeuvring is extreme and power is brokered between individuals. Organisations with a low power distance have disputes about the basis of power and may deviate from purpose, but they are more cohesive. Masculine organisations tend to be assertive and competitive, allowing them to take risks and be action-oriented. Feminine organisations favour cooperation and security, giving them more coherence but problems with delivery. In an individualistic organisation, everybody looks after themselves and the organisation's identity is based on individuals' own performance; such organisations cope with change projects through individuals manipulating rules informally in order to deliver the outcome that works for them. Collectivist organisations base their actions on working together, but again this can get in the way of delivery.

An important aspect is the way in which different organisations face up to their unknowns. High uncertainty-avoidance organisations require rationalisation of their unknowns, in the sense that they wish to be in control over them or at least reduce them to the minimum. This is most significant when decisions outside the normal decisions have to be made, as these involve acknowledging the unknown and making judgements with incomplete knowledge. These circumstances put stress on an organisation, induce emotion and can be destructive for the normal operation.

Thus, the informal organisation may not be negative; it may enable the overcoming of unknowns by creating acceptable solutions and explanations. Organisations, whatever their structure and purpose, have internal politics and different ways in which power is used and abused. However, these do create an apparent contradiction in the formal organisation and the view from the outside, which can be understood only by accepting the non-rational aspects of the informal organisation. Indeed, the informal organisation is the only arena in which contradictions are allowed to exist and ways of managing contradictions allowed to be developed. It is not possible to rationally manage a contradiction. In fact, rationality becomes its own contradiction in these circumstances. As an organisation tries to apply rationality, which cannot accommodate contradiction, then it creates a negative consequence, i.e. the action makes matters worse, and thus rationally managing contradictions creates a contradiction. If an organisation is forced to acknowledge the contradiction, then its rationality is wrong, which, in a world of rationality, is a contradiction. In fact, all organisations have problems in these areas – some more so than others, and some in some areas more than others. They are all, however, accommodated within the informal organisation.

3.5   PEOPLE IN ORGANISATIONS

By invoking the ideas of organisational culture and the informal organisation in the previous section, we are emphasising the importance of people in organisations. In our model in Figure 3.1, people are placed towards the unknown side; however, we do not see the unknown aspects as negative but as something to be managed and as something that can deliver positive outcomes. The formal pronouncements and artefacts of organisational life that are known are, to some extent, contrived; it is people that give organisations real identity and make purposeful things happen. The study of people in organisations variously covers psychology, social psychology, sociology and anthropology. This is another vast area of understanding that we reduce to three aspects: people are different, people have emotions and people have personal expectations. In the latter of these aspects, one critical issue is that people seek careers as individuals and take opportunities within their organisations in order to achieve this. We suggest that individuals' needs and emotions are exhibited in organisational behaviour (Huczynski & Buchanan, 2001) and vice versa, i.e. people exhibit emotion from the way in which organisations behave. It is these characteristics that we will need to appreciate in order to further understand clients as people, whether at rest or when building.

Although many aspects of peoples' behaviour in organisations are unknown, people do behave in purposeful ways, partly responding to external circumstances and partly driven by their internal thoughts.

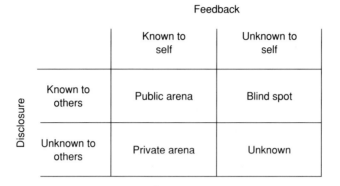

Figure 3.8   Johari window.

This means that they can follow plans like a machine but seek their own influence and rewards in situations. However, what people communicate may not be what they are thinking or feeling; thus, it is the meaning of their response that is unknown. This can be usefully modelled by the Johari window (Luft & Ingham, 1955), shown in Figure 3.8. This simple representation shows two dimensions – disclosure and feedback – which are essentially what individuals tell others and what others can tell individuals about their situation. Shared knowledge occupies the public arena; what is known to an individual but not to others is in the private arena. What is known to others but not to an individual is in the blind spot. And what is not known to an individual or others is in the unknown. This has been used by Barrett and Stanley (1999) to demonstrate the difficulties of getting a brief from a client; it is useful in understanding how various unknowns can be hidden from either the client or the industry.

People also have a view of their future, some of which will be in the private arena and so unknown to others. Thus, career aspirations, vendettas and manipulation fall into this arena. However, people's view of the future also contains emotion, of which individuals may not be aware or in control. This means that their reasoning may be in their bind spot but also may be in the unknown if it involves deep psychological compulsions. However, these Johari unknowns are categorised in our hidden or latent unknowns. The real future is contained in the unformed unknown, which is partly derived from a drive to succeed and partly from the developing conditions of the organisational situation.

People are different

People are different both in the way they see and in the way they experience the world (Cooper, 2002). This is more than their personality, which is their response to stimuli but also their way of thinking. No two people are the same; each person has different personalities, aptitudes and motivations. Differences in people are seldom acknowledged in the formal structure and processes of an organisation; these are seen as aspects to be suppressed or dealt with as unimportant noise disturbing the formal processes. The normal organisation learns to manage this formal idealised world and hides the accommodation of people, which gets subsumed into the hidden organisation introduced in the previous section.

There are a number of models of differences between people, e.g. learning styles (Honey & Mumford, 1992), Belbin team roles (Belbin, 2000) and the Myers–Briggs type indictor (MBTI) (Myers & McCaulley, 1985). We will use the MBTI as it is comprehensive and deals with individuals as well as group responses. The MBTI model of personality uses the three dimensions based on Jungian typology shown in Figure 3.9 (Huczynski & Buchanan, 2001) plus one other dimension labelled 'judging and perceptive'. The Jungian dimensions are *extrovert* and *introvert*, where people who prefer to direct their

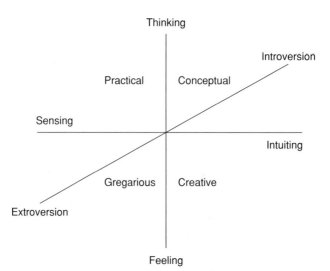

Figure 3.9   Jungian dimensions of personality.

energy to deal with people, things and situations are extroverts, and those who prefer to deal with ideas, information, explanations or beliefs are introverts; *sensing* and *intuitive*, where people who prefer to deal with facts and what they clearly know are sensing, and those who prefer to look into the unknown and generate new possibilities are intuitive; and *thinking* and *feeling*, where people who prefer to decide on the basis of objective logic, using an analytical and detached approach, are thinking types, and those who prefer to decide using values and/or personal beliefs are feeling types.

The last dimension of the MBTI – *judging* and *perceptive* (not shown in Figure 3.9) – is important in our model, as it considers the future, i.e. what individuals believe should be and can be brought about. People who prefer their lives to be planned, stable and organised are judging types, whereas those who prefer to maintain flexibility and respond to things as they arise are perceptive types.

The idea of formal management assumes that we can plan and then organise people to bring about the plan within the circumstances surrounding us. This is representative of judging types. These individuals believe that we can plan things into a scenario that is as accurate as possible and present this in a way that should be understood by everyone. If everyone followed the plan, then the project would be delivered to it and we would get what we wanted. Decision-making to judging types is a matter of collecting data and processing these logically through models (often numerical) in order to achieve the optimum answer. What is critical, then, is determining what is wanted at the beginning and following this through to the end. Deviations in the process are problematic and should be avoided. Judging types see problems as occurring because people do not follow the plan or the planning has not been done properly; therefore, they believe that we should spend time and resources to ensure that people do their jobs properly. In many ways, people are problems and need to be directed and checked continuously.

Alternatively, perceptual types believe that we cannot know what will happen in the future, and so detailed planning is a waste of time. It is a matter of working with what happens and making the best of this. Planning helps us to take the next step by giving us a reference point for action, but action cannot simply be directed by the plan.

These differences between people, particularly the distinctions between judging and perceiving and between sensing and intuiting, cause difficulties with coordinating the task, as groups of people work together. Different people see the problems in different ways and think that different things should be done. As organisations do not allow for differences but assume that everyone is operating the processes like a machine, then these problems cannot be acknowledged. This causes further problems, as it makes illegitimate the difficulties that are occurring. The situation arises most when groups have to undertake tasks that are outside the norm, when gaps, contradictions and unknowns arise. In the end, this creates further unknowns, some of which are hidden, some latent and, because of the uncertainty in knowing how people will react in the future, some unformed.

People have emotions

The emotional world is important in our study of clients. Emotions, however, do not occur only when a client decides to build but are present in an organisation at rest and held by its people (Fineman, 2003). The difficulty of understanding emotions is made worse by the fact that they are not regarded as legitimate in business, and thus they are not discussed. They can be regarded as part of the hidden unknown in any organisation. People and organisations learn to deal with emotions; this coping, however, resides at an informal level and a subconscious level, which again places it within our unknown. Clearly, this is a hidden unknown, as someone knows about it but is not expressing the fact or the consequences.

We think it is useful to divide people's emotions into aspirational emotions and consequential emotions. Aspirational emotions deal with the feelings individuals have about the future, i.e. those positive views that people desire and those negative outcomes that people fear. Consequential emotions are the feelings people have for events that have happened or are happening, and thus have a connection to immediate cause. The distinction is artificial, as feelings have a mixture of both at any one time. Consequential emotion can be caused by the gaps, contradictions and uncertainties in an organisation as well as by personality differences between workgroup members. These are often labelled as group processes (Huczynski & Buchanan, 2001). Thus, these gaps, contradictions and uncertainties not only disturb task coordination but also disturb individual and group motivation and responses to the task. Aspirational emotion comes from individuals' desires for themselves and their organisation. Thus, individuals' feelings are determined by their views on an organisation's potential to offer career advancement or a rewarding experience at work. In addition, these feelings can be shared when relating to the challenge of successfully meeting the organisational goals. Non-achievement of aspirations, or the fear of such, creates negative emotions.

There are a number of theories of emotions (Fineman, 2003), from biological determinism, whereby they are wired into genes, through psychodynamic, in which they are constructed in early childhood experiences, to social construction, whereby they are borrowed from cultural and organisational norms. Fineman (2003) indicates pragmatically that there is something to learn from each of these. The personality differences referred to above are based on the first two theories. The organisational culture referred to in Section 3.3 is based on the latter.

One theory that draws these different perspectives on emotions together is Kelly's (1955) personal construct theory (Bannister & Fransella, 1971). This focuses on individuals continually trying to understand and predict outcomes of the world that they inhabit. To do this, people invent models; it is these models or mental maps that are personal constructs. In Kelly's theory, the external world is real, but we experience it, appreciate it and pro-act on it through constructs; that is not the stimulus of the world but the interpretation of the stimulus. Each individual develops an idiosyncratic model of the world, leading to alternative constructions of reality to those held by others. Emotions (anxiety, guilt, threat, fear, hostility and aggressiveness) result as people meet situations that generate inconsistencies in their construct system. Anxiety is experienced when individuals' constructs do not represent a situation adequately. Guilt occurs when individuals act in a way that does not fit within their construct of themselves. Threat and fear relate to situations in which individuals' constructs may be invalidated in a fundamental or peripheral way, respectively. Hostility is the forced creation of a situation that allows individuals to reconfirm their constructs often as self-preservation. Aggressiveness is the emotion that individuals exhibit as they actively test out and explore a situation.

Individuals' constructs are formed from both their life experience and organisational experience. Thus, a group of people can share a construct (Bannister & Fransella, 1971) that determines that they react and have similar feelings in situations. This allows us to position an organisational emotion (Albrow, 1997) into the model. Organisations composed of sensing and judging types will feel unease at the unknowns in the organisation and suffer Cartesian anxiety (Thiry, 2001). This covers a response both to the ambiguity and to the uncertainty. Thus, organisations make sense of situations involving unknowns in their cultural and social context (Weick, 1995), partly to cope with the anxiety of the situation. This idea will be used later as we describe what happens when organisations decide to build. There are gaps, contradictions and unknowns in all organisations, and our point is that these are exaggerated when organisations build.

Effect of personal goals and expectations

Within any organisation, people have individual goals that are not aligned directly to the organisational goals. In particular, individuals may have aspirations to advance their position in the company or to leave the company for greater reward. Such aspirations cause people to be distracted from the organisational task and, in particular, to be less than willing to overcome unknowns, gaps and contradictions that might arise and that require extra work. The nature of these unknowns means that the person can legitimately say it was not part of the knowledge and processes of the organisation and so his or her actions were justified. There is a further problem that arises with desires to advance careers within an organisation and that involves competition between staff interested in gaining promotion. As well as being distracted, these members of staff may undermine each other's work and so cause problems for the task; also, teamwork for everyone becomes more difficult. For senior-level staff, such competition can cause a siege mentality between divisions that are meant to be cooperating.

Individuals have expectations and aspirations for themselves, which involve mental representations of the future. These produce a number of hope-casts and fear-casts (Miceli & Castelfranchi, 2002), which influence people's experience of what transpires in the future. This has a powerful influence on an individual's experience of an organisation and, collectively, on how the organisation is satisfied by future events. The experience results from two aspects: whether their forecasts have been validated and whether the result is good or bad for them. People get satisfaction from having their predictions validated. Emotions resulting from expected outcomes are less than those from unexpected outcomes, regardless of whether they are good or bad. Thus, the strongest negative emotions result from a prediction that some good event will occur, but it does not and the bad event occurs.

Thus, the failure in prediction exacerbates the bad outcome. The nature of this failure, against the norm of outcomes of such events, induces a further emotion as if the person was suffering something unfair.

> The stronger the hope-cast – that is, the more certain its implied prediction and the more important its implied goal – the stronger this sense of unfairness. In fact, an invalidated hope-cast can be called a *violated* one, as if some norm or prescription were implied somewhere.
>
> Miceli and Castelfranchi (2002)

Thus, it is difficult to produce a positive emotion satisfaction from planned activity. There is always something missing. In addition, any upset in the plans produces a much stronger emotion than the mere discomfort of the outcome. Miceli and Castelfranchi (2002) call this the 'negative power of expectations'.

3.6   CONCLUSION

The way in which clients think about the world is fundamental to the way that they see (and experience) the world and the way in which they make decisions within it. Clients' businesses do not work perfectly and clients do not have all the information needed in order to make decisions. We defined these gaps, contradictions and uncertainties as organisational unknowns; these put the client under stress in normal organisational operations. When the client comes to build, these issues are exaggerated. Thus, they are the starting point of engagement with construction and also the end point, in that they give clients their priorities and values. Understanding these helps us to determine what will satisfy and what will dissatisfy a client, even when a project is successful in building terms.

REFERENCES

Albrow, M. (1997) *Do Organizations Have Feelings?* London: Routledge.

Argyle, M. (1990) *The Psychology of Interpersonal Behaviour*. Harmondsworth: Penguin.

Argyris, C. (1993) *Knowledge for Action: A Guide to Overcoming Barriers to Organizational Change*. San Francisco, CA: Jossey Bass.

Atkinson, T. and Claxton, G. (2000) *The Intuitive Practitioner: On the Value of not Always Knowing What One is Doing*. Buckingham: Open University Press.

Bannister, D. and Fransella, F. (1971) *Inquiring Man: The Psychology of Personal Constructs*. London: Routledge.

Barrett, P. and Stanley, C. (1999) *Better Construction Briefing*. London: Blackwell Science.

Beer, S. (1985) *Diagnosing the System for Organizations*. Chichester: John Wiley & Sons.

Belbin, R.M. (2000) *Beyond the Team*. Oxford: Butterworth-Heinemann.

Boyd, D. and Kerr, E. (1998) An analysis of developer clients' perception of consultants. In *Proceeding's of the 14th Annual Conference of the Association of Researchers in Construction Management*, ed. W. Hughes. Reading, Vol. I, pp. 88–97.

Britt, S.H. (1979) *Psychological Principles of Marketing and Consumer Behavior*. Lexington, MA: D.C. Heath.

Bruner, J.S. (1958) Social psychology and perception. In *Readings in Social Psychology*, eds E.E. Maccoby, T.M. Newcomb and E.L. Hartley. New York: Holt Reinhardt and Winston.

Brunsson, N. (2000) *The Irrational Organization: Irrationality as a Basis for Organizational Action and Change*. Copenhagen: Copenhagen Business School Press.

Checkland, P. (1981) *Systems Thinking, Systems Practice*. Chichester: John Wiley & Sons.

Clark, P. (2000) *Organisations in Action: Competition Between Contexts*. London: Routledge.

Cooper, C. (2002) *Individual Differences*, 2nd edn. London: Arnold.

Crichton, C. (1966) *Interdependence and Uncertainty: A Study of the Building Industry*. London: Tavistock Publications.

Dawson, R. (2000) *Developing Knowledge-based Client Relationships: The Future of Professional Services*. Boston, MA: Butterworth-Heinemann.

Douglas, J., Field, G.A. and Tarpey, L. (1969) *Human Behavior in Marketing*. Columbus, OH: C.E. Merrill.

Drucker, P. (2002) *Management Challenges for the 21st Century*. Oxford: Butterworth-Heinemann.

Easterby-Smith, M., Thorpe, R. and Lowe, A. (1991) *Management Research and Introduction*. London: Sage.

Emery, F.E. and Trist, E.L. (1969) The causal texture of organizational environments. In *Systems Thinking*, ed. F.E. Emery. London: Penguin.

Espejo, R. and Harnden, R. (1989) *The Viable System Model: Interpretations and Applications of Stafford Beer's VSM*. Chichester: John Wiley & Sons.

Fineman, S. (2003) *Understanding Emotions at Work*. London: Sage.

Flood, R.L. and Carson, E.R. (1993) *Dealing with Complexity: Introduction to the Theory and Application of Systems Science*. New York: Plenum Press.

Goldratt, E.M., Schragenheim, E. and Ptak, C.A. (2000) *Necessary But Not Sufficient: A Theory of Constraints Business Novel*. Great Barrington, MA: North River Press

Hampden-Turner, C. (1984) *Gentlemen and Tradesmen*. London: Routledge and Kegan Paul.

Hay, J. (1993) *Working It Out at Work*. London: Sherwood.

Heirs, B. and Pehrson, G. (1982) *The Mind of the Organization*. New York: Harper & Row.

Herzberg, F. (1966) *Work and the Nature of Man*. London: Harper Collins.

Hill, R.W. and Hillier, T.J. (1977) *Organisational Buying Behaviour*. London: Macmillan.

Hofstede, G. (1994) *Cultures and Organizations: Software of the Mind – Intercultural Cooperation and its Importance for Survival*. London: McGraw-Hill.

Honey, P. and Mumford, A. (1992) *Manual of Learning Styles*. Maidenhead: Honey and Mumford Press.

Huczynski, A. and Buchanan, D. (2001) *Organizational Behaviour: An Introductory Text*, 4th edn. Harlow: Prentice Hall.

Hughes, O.E. (1998) *Public Management and Administration*, 2nd edn. Basingstoke: Macmillan.

Jacques, E. (1989) *Requisite Organization: The CEO's Guide to Creative Structure and Leadership*. Arlington, VA: Cason Hall.

Kelly, G.A. (1955) *The Psychology of Personal Constructs*. New York: Norton.

Kouzes, J.M. and Mico, P.R. (1970) Domain theory: an introduction to organisational behavior in human service organisations. *Journal of Applied Behavioural Science*, **15**(4), 449–469.

Lawrence, P. and Lorch, J. (1967) *Organisation and the Environment*. Boston, MA: Harvard Business School Press.

Luft, J. and Ingham, H. (1955) *The Johari Window: A Graphic Model for Interpersonal Relations*. Los Angeles, CA: University of California Extension Office.

McCabe, S. (2001) *Benchmarking in Construction*. Oxford: Blackwell.

McDermott, R.E., Mikulak, R.J. and Beauregard, M.R. (1996) *Basics of FMEA*. New York: Quality Resources.

McGregor, W. and Then, S.-S. (1999) *Facilities Management and the Business of Space*. London: Arnold.

Miceli, M. and Castelfranchi, C. (2002) The mind and the future: the (negative) power of expectations. *Theory and Psychology*, **12**(3), 335–366.

Miller, E.J. and Rice, A.K. (1967) *Systems of Organization: Task and Sentient Systems and Their Boundary Control*. London: Tavistock Publications.

Mintzberg, H. (1983) *Structure in Fives: Designing Effective Organisations*. Englewood Cliffs, NJ: Prentice Hall.

Morgan, G. (1997) *Images of Organizations*. London: Sage.

Myers, I.B. and McCaulley, M.H. (1985) *MBTI Manual: A Guide to the Development and Use of the Myers–Briggs Type Indicator*. Palo Alto, CA: Consulting Psychologists Press.

Nonaka, I. and Takeuchi, H. (1995) *The Knowledge-Creating Company: How Japanese Companies Create the Dynamics of Innovation*. Oxford: Oxford University Press.

Porter, M. (1985) *Competitive Advantage: Creating and Sustaining Superior Performance*. New York: Free Press.

Quinn, R.E. (1988) *Beyond Rational Management: Mastering the Paradoxes and Competing Demands of High Performance*. San Francisco, CA: Jossey-Bass.

Ranson, S. and Stewart, J. (1994) *Management in the Public Domain: Enabling the Learning Society*. New York: St Martin's Press.

Rush, R. (ed.) (1991) *The Building Systems Integration Handbook*. Oxford: Butterworth-Heinemann.

Simon, H.A. (1976) *Administrative Behavior*, 3rd edn. New York: Free Press.

Solley, C.M. and Murphy, G. (1960) *Development of the Perceptual World*. New York: Basic Books.

Stacey, R.D. (2001) *Complex Responsive Processes in Organizations*. London: Routledge.

Sveiby, K.E. (1997) *The New Organizational Wealth: Managing And Measuring Knowledge-based Assets*. San Francisco, CA: Berrett-Koehler.

Talbot, C. (2003) How the public sector got its contradictions: the tale of the paradoxical primate. *Human Nature Review*, **3**, 183–195.

Thiry, M. (2001) Sensemaking in value management practice. *International Journal of Project Management*, **19**, 71–77.

Thompson, J.D. (1967) *Organizations in Action*. New York: McGraw-Hill.

Walker, A. (2002) *Project Management in Construction*, 4th edn. Oxford: Blackwell Science.

Weick, K. (1995) *Sensemaking in Organisations*. London: Sage.

Williamson, O.E. (1975) *Markets and Hierarchies: Analysis and Antitrust Implications*. New York: Free Press.

Wilson, R.W. and Harsin, P. (1998) *Process Mastering: How to Establish and Document the Best Known Way to Do a Job*. New York: Quality Resources.

Winch, G.M. (2002) *Managing Construction Projects*. London: Blackwell.

Wright, S. (ed.) (1994) *Anthropology of Organisations*. London: Routledge.

4   The Client in Change

4.1   THE PROJECT MEANS AND ENDS

This chapter is about *change*, including creating it, managing it and experiencing it. Buildings are about change – change in the physical world and change in the organisational world of the client. Change involves a means of change and, at least, a concept of the end of change. Means have an instrumental and active character, whereas ends have a terminal and completed character.

Although this chapter is concerned with means and ends, it fundamentally involves the other dimension of knowledge and processes. We propose that all change situations have two routes: a known, formal, rational route and an unknown, informal, emotional route. Thus, we have four areas of concern: known ends, known means, unknown ends and unknown means (Figure 4.1).

Our intention is to highlight the role of the unknown means and ends in any situation of change to a greater extent than has been acknowledged before. Of course, the known rational route and the unknown informal route operate together, and it is only for the purposes of analysis that we are separating them. In fact, it is the interaction between them that causes some of our problems of understanding clients and of delivering successful projects and satisfied clients.

The philosophical aspects of this are important:

> The 'End' has initially only an ideal existence, and the Realised End – the actual outcome of the adopted Means – may be quite different from the abstract End for which the Means was adopted in the first place. Both Means and Ends are therefore processes which are in greater or lesser contradiction

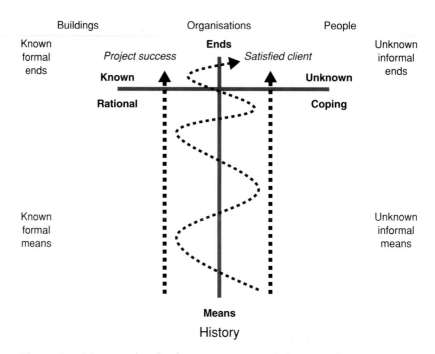

Figure 4.1   Means and ends of project success and client satisfaction.

with one another throughout their development – constituting a learning process of continual adjustment of both Means and Ends in the light of experience – until, at the completion of the process, Means and End merge in a form of life-activity, which is both its own End and its own Means.

Encyclopedia of Marxism (2005)

Thus, in reality, as we discuss later in this chapter, all ends are unknown at the beginning of a project, becoming known only as the project transpires through the application of the means. The end starts as an aspiration. The problem with the aspirational is that it is not formed completely, and certainly its satisfaction is unknown. It is also dependent on where the process of change starts; in this sense, the history of the organisation and its experience of building all come into the means that determine the ends. There are also issues about cause and effect. In such a complex interactive and dynamic situation, the identification of cause and effect in these circumstances is barely possible and, thus, is unknown (Flood & Carson, 1993). Our tools of knowing are inadequate, and so again our outcomes are unknown at the beginning of change.

We consider the values behind means and ends as one way of tying down agency in this process of change. We then consider how building involves organisational change in the client as a fundamental awareness of the book. The issue of how much we can know in this process of change is discussed next in order to establish that building and organisational change involves unformed unknowns that cannot be known beforehand. The emotions behind change are one of the unknowns, and we describe the way in which they are both created by the change and also influence change. The way in which change creates gaps and contradictions is then established, and this allows us to look at the consequences for the client organisation and for the project. In order to complete our analysis, we return to the interaction between the industry and clients, where we analyse the problems of engagement, in particular identifying how each can oscillate between a power and a dependency position, producing a dysfunctional process. We intentionally exaggerate the problems in order to make them have significance; in reality, clients and projects reach an accommodation that produces success more often than not. Chapter 3 and this chapter allow us to create a format to analyse client sectors in the chapters that follow.

4.2   MEANS AND ENDS AS VALUES

Rokeach (1973) states: 'A value is an enduring belief that a specific mode of conduct or end-state of existence is personally or socially preferable to an opposite or converse mode of conduct or end-state of existence.' Although value in the industry is identified mainly with money, several authors (Green, 1994; Hutton & Devonald, 1973; Kelly *et al.*, 2002) appreciate that values are what clients and the industry use to make decisions and take actions. This led to the 'Be Valuable' initiative (Saxon, 2006). Both means and ends connect to an organisational value system (Rokeach, 1973), which connects to the idea of an organisational culture introduced in the previous chapter. Means involve instrumental values and ends involve terminal values; examples are shown in Table 4.1. Values determine what we think of as good and bad and tend to have a bipolar character, i.e. every good value has an opposite bad value, which can change depending on circumstances. People and organisations determine their objectives on the basis of their values, search for suitable solutions, evaluate these solutions and finally make a choice. Satisfaction then is based on the achievement of desired end

Table 4.1   The instrumental values and terminal values of Rokeach (1973).

Instrumental values	Terminal values
Ambitious	A comfortable life
Broadminded	An exciting life
Capable	A world at peace
Cheerful	Equality
Clean	Freedom
Courageous	Happiness
Forgiving	National security
Helpful	Pleasure
Honest	Social recognition
Imaginative	Salvation
Independent	True friendship
Intellectual	Wisdom
Logical	A world of beauty
Loving	Family security
Obedient	Mature love
Polite	Self-respect
Responsible	A sense of accomplishment
Self-controlled	Inner harmony

values through a set of desired instrumental values. There is not a direct relationship between instrumental values and end values. Different instrumental values can contribute to one particular end value, while one instrumental value can also contribute to different end values.

In change situations, the instrumental values become of greater importance. Our means have a degree of clarity about them as we experience them immediately; in simple terms, our means are the relationships, coordination and administration of physical resources, people and organisations and the management of boundaries between them. All processes are means, but during change organisations need different processes compared with when they are 'at rest'; these different processes will seldom be easily stabilised. Thus, change exposes value conflicts between the past and the future, and between knowledge and the unknown. Our means are not only our methods and actions, but also our reasons; thus, the apparent simplicity of methods of action hides a number of much deeper concerns and involves our values in undertaking tasks. For clients and the industry, there has been a tendency to believe that the ends justify the means. Some of these means are determined by the external world, which has created established norms. These norms are set by the culture – organisational culture (see Section 3.4) embedded in the sector culture and sector culture set in the national culture. It is difficult, but not impossible, for organisations to work with a different value set than the norm. Unfortunately, this is a requirement of doing things differently. One difficulty is that within organisations, these new instrumental values can be held differently in different divisions or can be held within the project but again cannot be held outside the rest of the client organisation.

We use our model dimensions to present examples of values for change situations in our three achievements – buildings, organisations and people – and these are developed in Figures 4.2–4.5. It is not possible to detail everything in these figures, as all lists are inadequate summaries of values, which are context-dependent. In addition, the values presented are not independent. Thus, in any situation, the values operating can be different and can operate differently (Rokeach, 1973).

Values and buildings

The values surrounding building are shown in Figure 4.2. This is the world with which the construction industry is familiar. We have distinguished between the two routes of known rational means and ends, and unknown non-rational means and ends. The building is the central point of the project. Its physical reality, or its potential reality, within the project gives it an authority because of its substance. As we have said, this leads the industry and clients to believe that building is about building rather than organisational development. We fundamentally believe that technical building is both important and difficult in its own right, and the existence of knowledge and skills within the unique organisational setting of the construction industry pays testament to this. Indeed, skills in this area may be undervalued and underappreciated by clients. Of course, the technical delivery of a building is a necessary condition of satisfaction, but it is not sufficient by a long way. It is the way this interacts with and gets disturbed by the other areas that we need to understand and, in many cases, where our difficulties lie. There is a physical substance to buildings; therefore, in order to achieve the values of technical building, we need a model of what clients expect for this and what it is possible to get. We believe this is the normal way of dealing with clients and buildings, and it is addressed by various models of briefing (Barrett & Stanley, 1999). This is tangible, at least in the end, but buildings are a lot more than

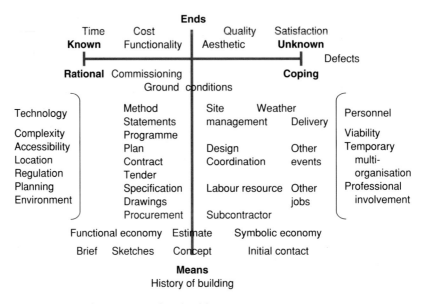

Figure 4.2   Values surrounding building.

their physical existence. Our industry is expert at delivering physical buildings: they have been trained to do it, they enjoy doing it, and the organisational systems have been evolved to make this happen. Whether this is satisfying for the client is another matter.

The known route has the familiar ends of time, cost and quality and the familiar formal means of the industry from concept, through briefing, procurement, design and construction. This is where the majority of studies to improve the construction industry have concentrated, occasionally drawing from the other route to modify practice. There are unknown aspects even of these familiar values. The technology of building can be seen as the driver of the known side and the personnel issues as the driver of the unknown side. Some familiar values, such as quality, we have placed more on the unknown side, as it is an area where equivocation is possible; the meaning depends on the circumstances involving the building, the organisation and the people that surround the end and the way in which this end was arrived at. Similarly, contract, as a means of change, formally structures arrangements, even if in the end it is indexical (Clegg, 1992) and open to interpretation, placing it partially in the unknown. However, although the subject of contract, the conflicts of getting the right personnel on to a job at the right time are a complex negotiation between other work that is being undertaken and the particular commitment to the job in focus. In the end, satisfaction requires fulfilment in both routes so that, for example, aesthetic appreciation may start in the formal briefing processes but be consolidated through empathetic interaction between people. Planning and building regulations are known technical values in this context, constraining the delivery. It also has an unknown technical component concerning the interpretation of what is acceptable in aesthetic or use terms. Planning, as development control, also has an intervention in the organisational value chart, but then it is on the unknown side.

Values and organisations

We have produced a values diagram for both public and private organisations, as they need to be seen differently; these are shown in Figures 4.3 and 4.4. The details in these figures are discussed in Sections 4.3, 4.4 and 4.6, as the organisational aspects are key to

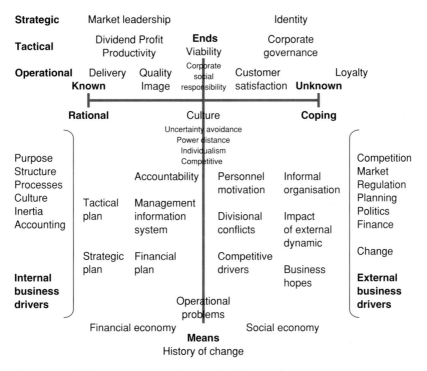

Figure 4.3   Private sector organisational values in change.

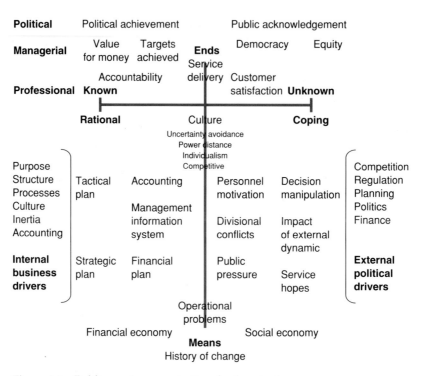

Figure 4.4 Public service organisational values in change.

satisfaction but also the route of many problems. Organisational values are set by the organisation at rest; the needs of this were described in Section 3.4. In organisational terms, end values tend to be concerned with profit, future existence, market leadership, industry respect, aesthetic beauty, identity and winning. The instrumental values of fairness, honesty, civility, loyalty, collective responsibility, efficiency and competitiveness are all set within norms of business of clients and the industry. Some of these are presented explicitly in mission statements or value statements, but there is a degree of unknown surrounding this, as there is a gap between espoused theories and theories in use (Argyris, 1993). In addition, not all of an organisation's values are articulated – many are latent. It is possible that there are inconsistencies, particularly as an organisation moves from stability to a change situation and its values are pressurised. A double standard can exist between the internal perspective and the external perspective. Thus, values are assumed to be known, but all organisations have unknown aspects because of interpretation, errors and interactions. Development planning regulations are part of the unknown, because they are open to negotiation, depending on the significance of impact of the organisational change on local economic development, and because development often exposes value conflicts in the local political arena.

Values and people

The values people hold as individuals colour their outlooks and determine their actions in situations; these are shown in Figure 4.5. This was introduced in Section 3.5. These values have a rational dimension in the sense that people have plans for development that are explicit and understandable in relation to the tasks of the organisation. Thus, organisations create career plans and review the performance of individuals partly to help the individual and partly to help the organisation. A person's location within an organisation, role and remuneration are all determined within this process. There is a growing agenda of work–life balance and new employment legislation, which gives individuals rights as part of their commitment to the organisation.

The non-rational unknown route, however, interacts continuously with this rational route. It is not possible to treat employees only as human resources; nor is it possible to assume that employees will act only as organisational resources. It is this emotional dimension that is significant during change when opportunities become available or losses are experienced. This is discussed in greater detail in Section 4.5.

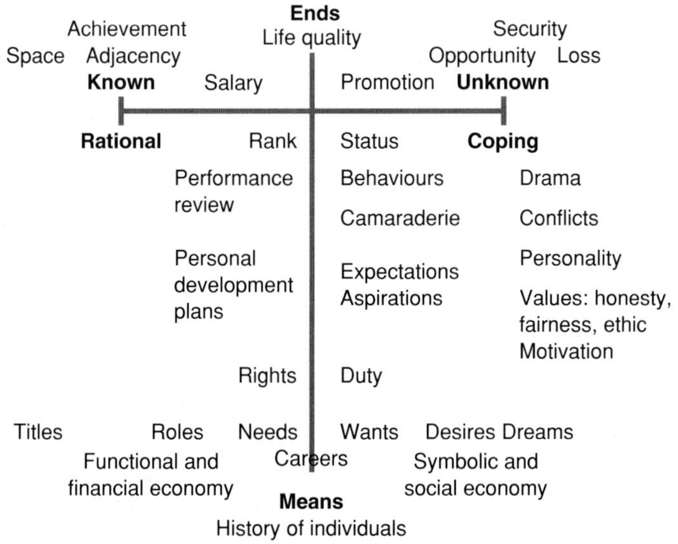

Figure 4.5   People's values during change.

4.3   BUILDING INVOLVES ORGANISATIONAL CHANGE IN THE CLIENT

In a building project, it is obvious that there is change involving the physical environment. A major concern of this book, however, is the fact that, equally importantly, there is change in the client organisation alongside a building project. This change starts before and continues after building. We believe that the absence of thought given to this causes some of the problems in the industry's engagement with clients and that it is something that needs to be managed. The client induces this change itself, and this brings us to look at the business strategy of clients and their capacity to accommodate change. Thus, we see building, as Boyd and Wild (1999) do, as a form of organisation development (OD).

The change that we are talking about is, like all business change, concerned with the client providing greater value for its business environment. The nature of that greater value depends on the nature of the client's business or service and the circumstances of the client's business environment. The greater value for an individual client building his or her own house could simply be the seeking of self-aggrandisement, and for a global company building a London office could be an increase in market share; the relation of the value to the building may not be obvious. The objective is not the building itself but the aggrandisement or increased market share. This book tries to explain how a building will help change the client in order for the client to achieve its objective. In organisational terms, this change may be concerned with business expansion or downsizing. The organisational aspects of this are clear, as are business projects in service improvement or the adoption of new technology. Other changes can be because of building obsolescence, which also involves organisational change associated with a new location or new facilities. The advent of building in the client's mind provides a sense of opportunity of what can be better for the organisation and its people, coupled with a sense of loss of what has been (Tannenbaum and Hanna, 1985). Thus, building requires people to change, organisational processes to change, and even organisational purpose to change.

A similar stance has been adopted by Winch (2002), although Winch continued to see the problem as managing construction projects. A wider approach was developed by Boddy and Buchanan (1992), who took generic project management explicitly as managing change. Indeed, many project-management authors see their key role as managing change, although there can be a confusion here, as change managers can see what they are managing as projects. We too might see a project the means of achieving the organisational change. However, much project management views itself as managing and integrating the tasks around technical building (Walker, 2002) rather than managing a wider system. In our sense, then, we see project management also as involving management of the client, similar to the network centrality role suggested by Lord *et al.* (1990) and Ibarra (1973). This is in no way a simple addition to the role; it is a fundamental repositioning of the role, with the management of the task being placed elsewhere in the system. We believe many project managers adopt this new role, although this reality may be hidden by convention of what they should be doing or what their fee agreement says they do. It is clear that it requires a new knowledge and power position, which Lovell (1993) started to explore.

A better source of ideas for our problem, rather than the construction industry literature, is the organisational change and behaviour literature. Organisational change is a major issue within management. Most literature on change is concerned with strategic business development, which deals with the fundamental purpose, structure and operation of the organisation. Sometimes change is deliberate, coming out of conscious reasoning to determine ends and conscious actions as the means (Iles & Sutherland, 2001). This type of change is called planned change or goal-directed change. Alternatively, other change appears spontaneously and is unplanned, and this is known as emergent change (Iles & Sutherland, 2001). The majority of the literature discusses goal-directed change, in the sense that a favoured outcome is set and the means of achieving it is devised. We follow Stacey's (1996) discussion and conceptualisation of change, as Stacey is one of the few authors to deal with the un-intentionality of change. The work of Morgan (1997) aligns with this and gives us a tool to analyse differences in the way organisations conceptualise change. Stacey (1996) sees goal-directed change as part of the rational model of strategic management. Stacey and many other authors identify that goal-directed change seldom produces the desired outcomes, and many projects fail because of problems of implementation. Stacey postulates other models of change, which are useful in our conceptualisation of the change induced by building.

The distinction between goal-directed and emergent change is important. The former involves rational planning as a means to achieve the goals at the end. This falls within the known arena, with known ends and using known means to achieve them. It would appear from the rational planning perspective that any building project should be a goal-directed change and that this is knowable. In addition, strategic management is a way of creating goal-directed ends of change and the plan is the means for achieving them in organisations. Again, it appears that these are all knowable. However, change can be emergent rather than planned in two ways: the assumptions used for the plan are not completely valid, and the external world can change, making the purpose of the change less valid. Thus, all change is non-linear and contains both goal-directed and emergent aspects and is, to some extent, unknowable.

The issues surrounding change concern upsetting the stable system of client operation. Stacey (1996) identifies three levels of change: closed change, contained change and open-ended change. Closed change can be regarded as stretching the current optimised system. The idea here is that there is a degree of linearity in the change and that the optimum is returned to, but at a different level of throughput. The linearity refers to the fact that all that is required is the same multiplier throughout the system. Although the system is disturbed, this is limited. It is the fact that it is knowable and predictable and so controllable that allows everyone to understand it.

Contained change involves a significant shift in the system; the overall system, however, stays the same. There is agreement on the whole but disagreement on the parts. There is a degree of non-linearity in this, in that some parts of the system may be reduced whereas others may be expanded. Some bits are knowable and others may be forecast; for example, repetitive tasks may be automated, thus requiring fewer staff. It is possible to predict and control the overall system, but it is less clear how the parts will develop.

Open-ended change involves a shift to a new system. There may be considerable disagreement about what the system should be or even why there should be a change. References to the past system may be not relevant, and therefore very little is knowable. It is not possible to predict the operation of the system and, indeed, it is probably known that bits will not work. Thus, control is more difficult and may induce problems itself. If more than one aspect is working suboptimally, then it is difficult to know what to control.

In a client that is building, then open-ended change is occurring through the two routes (known or goal-directed, and unknown or

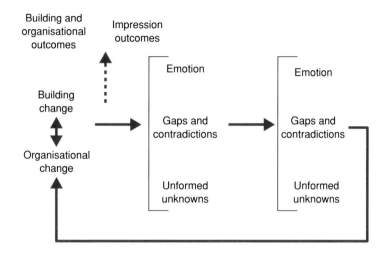

Figure 4.6 Interacting consequences of change.

emergent), and these routes interact. In our model of the client, we need to understand the knowledge the client has of its situation, the level of change that the act of building is inducing, and whether the client is managing the unknowns suitably. The consequences of this change are operational in creating gaps and contradictions, emotional for the people, and unpredictable, as some change is emergent during the process. Figure 4.6 shows the interaction between these consequences of change. It is difficult to separate client change from client internal gaps and contradictions, from the emotion of change and the unformed unknowns of change. Each induces the other, and this can escalate to a project being a catastrophic failure. The outcome, when an end is reached, is constituted partly by the end building, partly by the end organisational state and partly by the impression of the process.

Although we discuss separately the unformed, the gaps and contradictions, and emotions, they are completely interdependent.

4.4 BUILDING INVOLVES UNKNOWNS THAT ARE UNFORMED

In Section 3.3, we defined three levels of unknown: hidden, latent and unformed. Hidden involves knowledge not being revealed, latent involves a non-realisation of connection or existence, and unformed means unknowns that have not come into being yet. In reality, all ends are unformed at the beginning of a project, becoming known only as the project transpires through the application of the means. The project may even be cancelled during the process. These emergent or unformed aspects of change involve some latent issues, which could be discovered with more detailed attention. It is the unformed unknowns where no amount of work will reveal their nature that are really problematic. Building has a high degree of emergence and, thus, a high degree of unformed unknowns. The significance of hidden and latent unknowns is an unformed unknown. In particular, client satisfaction at the end relates to the client's aspirations; however, the aspirational is not formed completely at the beginning of the project but is created during it.

The high degree of unknowns during building is a result of the complexity of the situation and the dynamics of change. The Tavistock studies of the building industry identified this as uncertainty and inter-dependence (Crichton, 1966). First, there is change in both the client and the building project; these interact, causing confusion of purpose and opportunities for manipulation. Decisions in one domain can be justified by problems in the other. Problems in one domain can be blamed on issues in the other. Alternative objectives can be surfaced from organisa-tional contradictions or from personal agendas, which can be presented as essential needs. This causes changes, i.e. emergence in specification and in aspiration. This is discussed further in Section 4.6.

Second, the assumptions about the world are not accurate, and thus organisational business plans and building procurement and design plans do not work completely. Although the organisational world may be quite bounded and understood, such that a degree of goal-directed control is valid, during change this understanding will be pressurised and deviations will occur. For example, the assumption about the use of space normally to undertake a task may be valid and used in design, but for an improved service with the use of new equipment then the space required may need to be changed when more details of the new equipment are received. There is a multiplicity of small items, which together form a greater problem.

The assumptions about the world of construction are much less well bounded, as it is temporary, involves disparate parties and takes place in a new location. This refers to the temporary multi-organisation character of design and construction (Stringer, 1967; Walker, 2002). Again, the degree of change means that understanding of even basic processes is pressurised, and the project continually has to be pulled back on to course or the course redefined. For example, the assumptions about how long an activity takes may be valid only if other project parties fit in perfectly; this degree of coordination across a temporary multi-organisation when people are present only for this event is extremely difficult for time-planning but also even for task-planning. The degree of management required to undertake this is only possible economically on very important aspects, as it involves continual double-checking of arrangements to ensure communications and acceptance and having emergency resource at hand. Even in design, which has a virtual shared communications medium of drawings, the coordination of designed parts from one organisation with those of another requires their interdependence to be understood completely. The concept of reciprocal interdependence (Thompson, 1967) explains the nature of these unknowns. Again, the significance of deviations in understandings is difficult to appreciate among other coordination problems, and thus they emerge as the project develops, sometimes being appreciated only at the point of physical construction.

This aspect of inaccurate assumptions can be partly rationalised as risk, and there are many tools available to evaluate it and to make decisions with risk. These risk tools are most successful when they deal with aspects of the unknown that are hidden or latent, i.e. that yield to probability. This can help us to understand sensitivities to change and to different decisions. However, risk and risk tools do not cover unformed unknowns (Casti, 1993), and so we are left with a residual unpredictable change.

Third, the external world changes, and this interferes with the relevance of the objectives in both the organisational change and the building change. This can be strategic, for example an economic down-turn putting into question the business reason for undertaking the change, or operational, for example another construction project in the region means that labour is hard to obtain.

Fourth, means are formed by means rather than by design. In a situation of change, some of the means emerge, and thus there is a chain reaction of unformed outcomes. Ideas come out of the unknown because they are not completely organised and are hidden by indi-viduals' hopes and fears. They move into the known through a process of meetings, studies and reports that are part of the normal means (processes) of an organisation. The means of the organisation for dealing with the idea may not be sufficient, and therefore new means may be part of the idea, for instance appointing an architect. The sequence of means is important as, for example, the outcome might be different if a quantity surveyor was approached first. Thus, the overall construction process that moves from the brief to specification to construction is, in fact, non-linear. The conventional tools cannot manage this non-linearity and degree of unknown. For example, contracts merely shift the impact of the unknown, which all parties are seeking to avoid, and so contracts can become part of the problem (Clegg, 1992). An amalgam of old and new means is produced; every new means adds complexity to a situation, thus making it more unknown (Leaman & Bordass, 1993). There is also a degree of game-playing, which, although merely a hidden unknown, induces unformed unknowns as other players react to the gaming situation (Hargreaves-Heap & Varoufaki 2004). As change advances, it is transformed by the nature of the end objectives, the means for working on them and the external conditions that were used to validate the idea.

Fifth, there is a conflict between organisational culture and these uncertainties of the future. Simultaneously, known rational processes and unknown non-rational coping are taking place, which interact and cause unformed outcomes. The rational goal route sets up prediction expectations, both about the building and about the organisation, in relation to the plans. The idea that the problems in this section can be brought under control by using concepts of risk is one such route. As reality has to accommodate emergence, the deviations from the plans and attitudes to these are emergent features themselves. Individuals' personalities affect the way they experience this. Invoking the negative effect of expectation, as explained in Section 3.5, these deviations, when they create losses for people, are seen as unfair. This feeling induces negative behaviours that are part of the means and cause outcomes that were not planned.

The development of these unformed unknowns can be understood by considering how the plans and aspirations develop over time. Simplified representations of this are shown in Figures 4.7 and 4.8. Figure 4.7 concerns what we know about the relationship between plans, objectives, formal outcome and reality. Figure 4.8 concerns what we know about aspirations, human engagement and fulfilment and reality. In each case, reality differs from these planned and aspirational routes and also changes itself. At the start, both the plan and the aspirations have greater knowledge than reality, which is completely unformed. In the end, reality is completely formed, but our plans

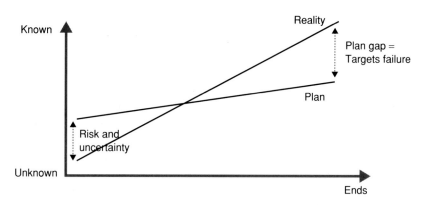

Figure 4.7   Known and unknown development of plans against reality.

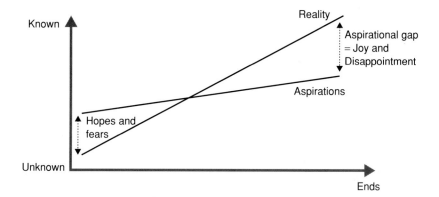

Figure 4.8   Known and unknown development of the aspirational against reality.

and aspirations still have unknown aspects. We have represented the changes as linear for clarity but in practice they will be non-linear.

The results in the end are plan gaps and aspirational gaps, which are partly unknown but are critical to the client's perceptions of satisfaction.

4.5   EMOTION OF CHANGE

The world of business assumes an unemotional rational process of relationships and decision-making. As we discussed in Section 3.6, people and organisations have emotions, even in a steady-state organisation. In change, more major emotions can emerge because norms are upset, opportunities for personal advantage are made available, and fears of personal loss are present (Fineman, 2003). What makes emotion so unavoidable in the change involved with building is that the change is substantive, involving major resources and significant effects. These emotions we regard as part of the unknown and coping side; emotional effects are always hidden or latent, as they could be found out with appropriate processes, but the consequences are unformed. Emotions are also part of the known dimension in that we know when we have upset someone or even that we are going to upset someone, and this knowledge can in itself be emotional.

The basis of emotions was introduced in Section 3.5, where we split emotions into aspirational and consequential. This focused on the emotion behind individuals' expectations and careers. We also alluded to the fact that organisations experience emotion (Albrow, 1997), whereby there is a shared response to an experience or an opportunity. During change, the rupture in the norm causes the information about the future to be both ambiguous and uncertain, heightening the Cartesian anxiety (Thiry, 2001). Groups partly create this anxiety by pointing out the potential problems and searching for answers (Lawrence, 1999) but also partly alleviate the anxiety by having a sense of shared problems (Cardonna, 1999). In order to accommodate change, people's constructs have to be revised, which itself causes emotion (Bannister & Fransella, 1971). There are, however, aspects of denial, particularly if change is seen as a loss (Tannenbaum & Hanna, 1985). The unformed unknowns disturb this situation more by challenging the security of the rational plan. The deviations from the plan in reality suggest that the plan is inadequate and that this course of action may be out of control (Obholzer, 1999).

Although organisational change and building can induce anxiety, they can also provide the potential for great elation and a sense of opportunity. For some individuals, then, the change induces a greater drive and even a competitiveness to succeed. This draws us back to the fact that people are different and will see situations of change differently. The result may be a polarisation within an organisation between those who see opportunity and those that see loss in change. During uncertainty, individuals can manipulate interpretations and decision-making in order to facilitate their gain. This can be in career terms or for resources within a new facility. The sense of this happening induces further emotion in those disadvantaged by the change and creates further polarisation.

Those promoting the change project have to overcome trepidation in other parties in order to advance their position. To do this, they need to exaggerate the benefits and downplay any dis-benefits or problems. This selling of the idea makes the expectations higher and the possibility of deviations more likely. There is much at stake on both sides, and this induces further emotion from the fear of being wrong. This is exacerbated by the fact that procuring a building is actually a buying decision, and all buying decisions involve emotions. It is not only a buying decision, however; also, people may not know what they are buying or know how the future will transpire as to whether what they will buy relates to what they will find. This is normally referred to as risk; as mentioned previously, there are substantial literature and tools for evaluating and handling unknown risk. Risk has an unknown psychological aspect, however, that involves fear of the decision, which affects the ability to make the decision.

Although satisfaction can be defined purely within mechanistic and functional domains, there is a hidden dimension that must also be satisfied; this is the most difficult dimension to determine and to achieve. The emotional aspects of satisfaction involve the meeting of expectations and aspirations. In situations of change, the aspirations change with the developing project, as shown in Figure 4.8. There will always be a deviation between the aspiration and the end reality. The idea of satisfaction of expectation has to work against the negative power of expectations (Miceli & Castelfranchi, 2002), where any unexpected loss has a greater emotional impact than an equivalent planned gain. The deviation from the aspiration has to be made sense of, and it is this that will determine satisfaction (Weick, 1995). This sense can be positive or negative. It is created within the organisational group setting, which may have different perspectives across the organisation. The organisational satisfaction is based on the negotiation of the sense of the deviation in aspirations across the cultural and the political domain.

4.6 CHANGE CREATES GAPS AND CONTRADICTIONS

In Section 3.4, we considered how organisations have structural and process unknowns during normal operation. These are overcome in normal operation by experience and by the coping behaviour created by the informal organisation making it happen. Our contention is that in situations of change, these gaps and contradictions cannot be overcome so easily, and further gaps and contradictions appear. This challenges the constitution of the organisation, both for the justification of the change and for its normal operation.

As we mentioned in the previous section, the change associated with building is substantive. The idea of changing the organisation has a rollercoaster life: the commitment to it moves up and down. Most ideas never come to fruition but are rejected during the processes of evaluation or implementation. The idea for strategic change is likely to be challenged inside the organisation and changed before it ever gets into an executable plan. Strategic change, which might lead to building, generally involves more than one organisational division. The location in the organisation of the original idea of the change is important, as are the locations of rejected ideas. Organisational behaviour indicates that the idea will tend to advantage the division of its inception. However, there may be other divisions that are disadvantaged by the idea or at least have their ideas displaced by this one. This tension between divisions is present at the start. The client organisation is not unitary, and prior events within the client system affect the current conduct of each of the divisions. Cherns & Bryant (1984) explored the competition for scarce investment resources between different interests within client organisations and how this generated value conflicts involving power and manipulation of declared risks and costs. Power accrues to the 'winners', but they can become hostages to fortune in relation to the emerging success or problems of the project. The hostages include personal reputations and professional and functional interests, which focus differentially on the limits of time, cost, quality, etc. The location of the original idea is part of the means but only defines the ends from this perspective.

The controller of resources is an important player in the advancement of a change project, but he or she has their own objectives, which may be threatened by this. For the controller, the idea needs to be promoted in the rational route, which involves creating a business plan and feasibility studies that justify the course of action quantitatively against the use of resources. This rational promotion of the idea is part of the means. These plans also include some means of achieving the idea, which could involve a building solution. The selling probably requires looking favourably at the circumstances that enable the idea, saying the ends can be achieved easier and with fewer resources, and exaggerating the benefits and ends. Thus, a tension is set up between the promoters and the opposers, where each is demonstrating rationally the correctness of their position. The problem is made worse because it is accompanied by the emotional behaviour described earlier.

This tension between functional divisions is evident in both public and private organisations. The public sector may have a more procedurally-oriented approach to coordination. The use of procedures to make decisions can become extremely slow and cause delays in the development of a project. The procedures also provide an opportunity for disaffected divisions to cause problems for the project. This urgency on one side and resistance on the other presents both task and emotional difficulties, which can produce further unformed unknowns.

There is also a tension in level: in the private sector between strategy, tactics and operations, and in the public sector between political, managerial and professional. These tensions can be seen as problems of objectives, both in creating them without contradictions between them and in maintaining them against change. The contradictions are more evident in the public sector (Talbot, 2003), as the difficulties of working with value conflicts between external stakeholders are exposed. The private sector is not immune from tension in levels, and these focus mainly on conflicts between short-term and long-term objectives. These contradictions and conflicts are then mirrored within different parts of the project organisation, which cause gaps to be exposed.

When things go wrong, i.e. when deviations between reality and plans cannot be hidden, then emotional actions are exaggerated. A list of potential causes of these project unknowns is given in Table 4.2. The process of rational planning is used to find reasons for these unknowns and to identify the means that are not delivering what is wanted. There is a temptation for organisations to regard it all as tardiness, assuming that the plans were correct and that it is the implementation of them that has not been performed properly. In reality, it is difficult to distinguish between what is genuinely unknown and what has been caused by tardiness. However, at this time the blame culture induces defensiveness in all business dealings, in case it is shown that a division or an individual is at fault. In addition, there is a reduction in communications generally in order to avoid giving away information that might show fault. Overall, whether in the client or in the project, such action induces further unknown consequences.

Table 4.2   Project unknowns leading to deviations from plans.

Location	Potential causes of unknowns
Temporary multiple organisation	Different people and organisations being brought together for the first time
Task fragmentation	Communication and coordination of task across professional and trade divisions
Industry issues	Changes in regulation, legislation and overall performance create reference point for projects
Regional issues	Shortage of suppliers and skill resources
Physical issues	Unknowns in ground, material, material compatibility and the weather
Organisational issues	Material supply, labour, contract, coordination and bankruptcy
People issues	Careers, personal conflicts, work pressures, illness, accidents
Events	Chains of consequences

This situation can be induced not only by the project but also by the client. When the reality of a financial situation is revealed or external conditions change, then a cutting of overall project costs may need to be undertaken. Alternatively, there may be a desire to open a facility earlier than planned in order to make it add value sooner. This causes problems both within the client and within the project. Attitudes to this additional change may re-ignite structural and process problems in the client organisation. These problems may be in the 'blind spot' (see Section 3.5) of the client, where further problems occur due to their undiscussable nature. The effect ricochets around the client and on to the project, causing all parties to become more defensive and untrusting. Boyd and Wild (1999) used the ideas of over- and under-boundedness (Alderfer, 1979) to describe these phenomena, and this will be used in Section 11.2 of the toolkit as an indicator of action that is required to assist the client.

4.7   MEANS AND ENDS OF ENGAGEMENT

We are often told that building is not difficult. This misses the point, however. Building is difficult when we look at the complexity of the

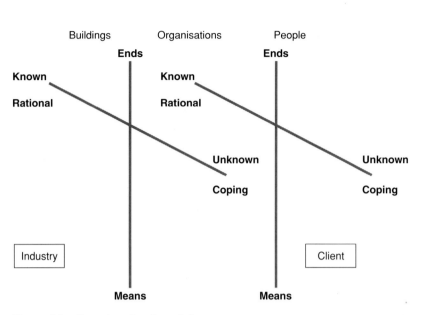

Figure 4.9 Complex situation of change.

issues demonstrated by our model. It is the physical building itself that appears tractable, and only in the rational route. We also know that the industry can solve most problems and, indeed, provide most dreams but this costs time and money. The dreams are to some extent non-rational, but the costs and time are rational. The industry's ability to integrate these, and the occasions when this has been done success-fully, drives others to believe that the industry can do it every time. This ability to have command over our physical world is compulsive; it enters us into the emotional world, such that if we think and know we can do it, then we are forced to ask why we do not do it. The industry and clients seldom see the implications of this.

An overview of the complexity of the situation of engagement is shown in Figure 4.9. The industry and the client are different entities and see the world differently (see Section 3.2); however, they both have known and unknown dimensions and experience means to deliver ends. Change is occurring in the client organisation, in the building project, and in the industry-base organisations and the temporary multi-organisation. The complexity comes from the interdependence of the three situations of change through the interacting of rational and non-rational means and ends. This is referred to by Clark (2000) as a competition between contexts.

The change in the building project and the change in the organisation involve different sets of means and different sets of ends. These were presented in Figure 4.2 for the building project and in Figures 4.3 and 4.4 for organisational change. The unknowns in these changes are significant, in particular because there is a high degree of unformed unknowns (see Section 4.4). These unformed unknowns are induced by changes in the external world, confusions of purpose about the organisations and the building, inaccurate assumptions about the world as it changes, and the emergent nature of means.

Other unknowns are present because of the formal differentiation in the client organisation (see Section 3.3). This differentiation is both vertical and horizontal, producing organisational islands, such that there are gaps and contradictions in the formal knowledge and formal processes preventing these from working effectively (see Section 3.4). In the normal operation of the client, these gaps and contradictions are overcome by hidden informal means; during change, however, these means are not effective, and the organisation has to deal with overcoming these problems. In addition, the organisational islands have their own objectives for survival, and these may be in conflict with the change objectives and the objectives of other organisational islands.

The response to these unformed unknowns, both in the project and in the organisation, can induce further unformed unknowns. The response is determined by organisational culture (see Section 3.4); if this abhors such uncertainties, such that deviations from plans threaten the organisation (e.g. public-sector organisations), then there is a move to allocate blame through the rational route. This creates defensiveness and dysfunctional communications throughout the organisation and project and so induces further unknowns. Formal management is used for the known project change and possibly for the known organisational change; there is little management of the unknown changes, however, or, most significantly, of the interaction between the project and organisational change.

This organisational and project change towards ends creates various gaps between reality and rational plans and between reality and aspirations. These gaps are themselves problematic. These are shown for the organisation in Figure 4.10 and for the project in Figure 4.11. These diagrams slot into the overall representation shown in Figure 4.9, which is given in complete form in Figure 4.12. The client's perception of these gaps is the basis of client satisfaction from the overall enterprise of change.

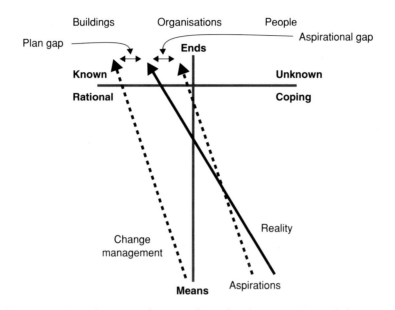

Figure 4.10   Development of gaps in the end in the organisational change.

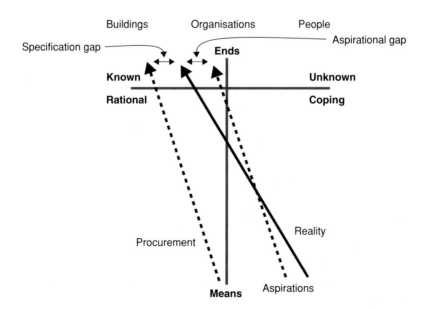

Figure 4.11   Development of gaps in the end in the project change.

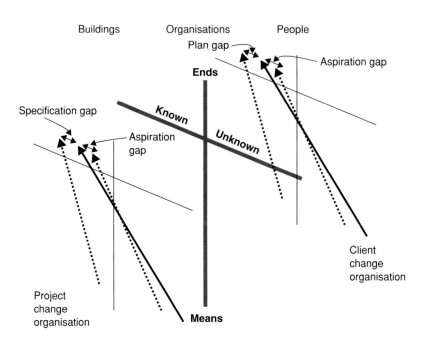

Figure 4.12   Picture of change.

These diagrams represent the idea that there are known rational and unknown non-rational routes in the change. These routes are conceptual in that they are in the minds of participants. The idea that there is a difference between conceptual routes and reality was introduced in Figures 4.7 and 4.8 for the unknown routes; Figures 4.10 and 4.11 have the addition of the known rational routes in the change. Reality moves from being virtually unknown at the beginning to being well known in the end. Reality may not be known completely in the end because we cannot see everything. The known rational conceptual routes involve plans and declared processes. This is known more than reality at the beginning of the project. The assumption is that the means of change will lead reality to be what the plan is. This will not be the case completely, however, and so we are left with a business plan gap for the organisational change and a specification gap for the building change.

At the same time, there are conceptual unknown non-rational routes involving the developing aspirations and expectations of the organisational change and building change. The deviation between these and reality at the beginning accommodates the hopes and fears of people and the organisation. The aspirational gap at the end involves the missed opportunities and unexpected advantages that the change has brought.

These four gaps have different characters and are seen differently by different parts of the client organisation and the industry. Gaps in the known route are more concrete and often can be quantified, such that they can be analysed, have blame attached or be justified. In the project domain, this known gap is the basis of claims, expert advice and legal jurisdiction. In the organisational domain, it re-ignites internal contradictions between organisational islands and between promoters and opposers of the change. These end-play situations colour the final outcome. These plan, specification and aspirational gaps invoke the negative power of expectations and as such create negative emotions that are difficult to overcome. The question is then whether the overall benefits are sufficient to overcome the gaps.

As stated previously, although the known project change and, possibly, the known organisational change are managed, there is little management of the unknown changes and, most significantly, the interaction between the project and the organisational change, which means that it is difficult to identify the aspect that is causing dissatisfaction. Problems in one area can cause problems in another area or can be identified as causing problems from the unknown non-rational route. No formal documentation can control or protect organisations from these volatile situations. It is not possible to rationally manage such contradictions. Rationality becomes its own contradiction in these circumstances. As organisations try to apply rationality, which cannot accommodate contradiction, they create a negative consequence, i.e. their action makes matters worse, and rationally managing contradictions creates a contradiction. Either the organisation is forced to say that the formal management is working, which is a contradiction because it is not working, or the organisation is forced to say that its rationality is wrong, which, in a world of rationality, is a contradiction.

Stacey (1996) provides us with insight into the problems of decision-making in those situations of change depending on the degree of uncertainty surrounding the situation and on the degree of agreement on what should be done. His analysis is shown in Figure 4.13. In this diagram, we see behaviours in response to the complex interactions and induced contradictions that are characteristic of the behaviours in the industry and with the client's engagement with the industry.

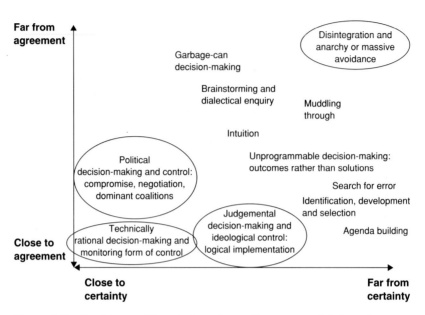

Figure 4.13   Decision-making strategy depending on uncertainty and agreement. Adapted from R. Stacey *Strategic Management and Organisational Dynamics*, 2nd edn. 1996. © Pearson Education Ltd. Reproduced with permission.

Industry response

Current construction management theory does not acknowledge these wider problems of change. As Boyd and Wild (1999) state, the conventional wisdom sees construction as building and views this in a rational process flow manner. What these conventional models and assumptions have as their basis is a stable environment in which thoughts flow into goals, which flow into actions, which in turn flow into outcomes. This is a systems engineering model and is extremely useful for improving systems that do not change either physically or in the mind. Current construction management theory delivers approaches to the strategic development of such stable systems. This may apply at an abstract level in construction on an industry basis, but not at an individual project basis or even a programme of projects basis. The conventional wisdom then is misapplied. It does not mean it is wrong or that the industry cannot learn from it, but one of the fundamental aspects of projects and programmes is omitted. Thus, in order to understand the client and the difficulties of the interaction between client and industry, this theory may not be helpful; indeed, it may actually be unhelpful, as it hides or does not see the fundamental problems of rational and non-rational change in both the client organisation and the project. Thus, working on improving the known rational route of the building change, such as finding better contracts, undertaking better briefing or visualising the project better, works only on part of the problem. The problem of the conventional is that it does not grasp the real problem: it idealises the situation and tries to maintain that incompetence is the cause of the ideal not being achieved.

Some of the current approaches to managing the situation between client and industry may benefit the situation. Techniques such as value management, partnering and framework agreements enable a deeper understanding of the joint situation to be established, both at a formal level and at an informal level, which can access the unknowns. These techniques then allow the industry to work towards reducing or managing the plan, specification and aspirational gaps identified above.

Problems of fragmentation and pairing

The problem of fragmentation in the industry makes the engagement problematic. This situation is represented on the left-hand side of Figure 4.14. The technical difficulties of buildings have resulted in

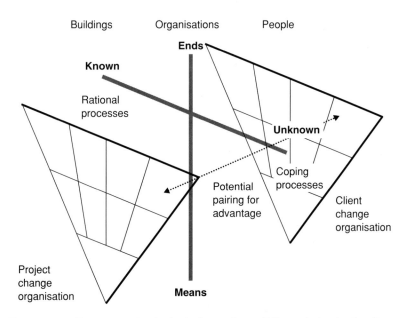

Figure 4.14 Fragmentation in the industry faces differentiation in the client.

fragmented entities with specialised functions. In Section 3.2, we identified that the industry saw the project differently from the client. We are now adding a further complexity, because within the industry specialised functions see the project differently; of course they need to, otherwise there would be no point in having specialised functions, which requires them to optimise their ability to deliver in this specialist area. This is one of the reasons why these specialisms have evolved. It was certainly easier when one person could conceive and undertake everything, whether as master mason or architect, but this has not been possible for a long time. We are left, therefore, with the perverseness of the fragmentation, both physically and organisationally, which we need to draw together in order to create a whole (Groak, 1992; Rush, 1991).

The client is also not unitary and has organisational islands that have different objectives towards the project. Thus, we have the complete situation represented in Figure 4.14, where facing each other are a differentiated client, with different values in its divisions, and a fragmented industry, with different values in its divisions. There is a tendency for parties across the divide to identify others with similar values and to feel an empathy that causes them to relate together. As the project transpires, client divisions negotiate for outcomes that meet their needs; when these needs are held strongly, there is the option to utilise this empathetic relationship in order to seek advantage. These informal relationships cross from client to industry and circumvent the formal processes. This pairing in the face of uncertainty (Bion, 1961) is most evident when there is a degree of disunity in the client and when the value conflicts are highest, for example surrounding the political nature of public service. It is problematic as it upsets the formal route and may cause distrust in the informal route.

We cannot, in fact, separate building from organisations and people. We can force organisations and people into a building engineering model, or they can accept a building engineering model. This all involves organisations and people. Therefore, even if there are pure building solutions, they do not manifest themselves without organisations and people. Thinking about them as independent can become part of the problem, as it apparently eliminates the need to manage the client organisation and people. For the industry, believing in only a building solution means it does not have to bother with the client's conflicting needs and their conflicting resourcing. For the client, believing in only a building solution removes these conflicting needs, but in a way that is more psychodynamic and involves deeper avoidance and projection.

Indeed, the client may have a hidden dream that if it concentrates on the building, this will overcome its organisational problems and people will be subsumed by the structure and scale of the physicality of the building. Facing these anxieties is unpleasant, and it would appear to be a lot better to believe in the building engineering view of the world. But then all have to accept the consequences, and the overall emotional world of building is unlikely to accommodate this. Thus, avoidance of the organisational sociological and individual psychological aspects of building (on both sides) is one major part of the construction industry's problem.

Power and dependency

The situation described above creates a basic contradiction in the relationship between the client and the industry. This is embedded in the definition of the client revealed in Section 1.3 between the client as customer or as a ward. The relationship is set as one of an individual who buys a service from another. This relationship, by its nature, ascribes power to the buyer. The seller is then dependent on the buyer for its reward for providing the service. However, because the buyer is unable to undertake the service itself, it is beholden to the seller for its service. This gives the seller power in the relationship, and the buyer is dependent on the seller. Such a problem is commonplace within professional relationships where there is a degree of mutual existence (Macdonald, 1995), but it is more problematic in situations, such as construction, where there is a clear divide between the parties. Indeed, the history of client–construction relationships has been one of master and servant; this class divide is evident in some of the constitution of the formal approaches of engagement as well as in the non-rational approaches. In the past, the master–servant relationship was mediated by an architect, whose role included being of an acceptable class to talk to both clients and servants (Hampden-Turner, 1984; Macdonald, 1995).

Although the class divide is no longer an issue, the emotional position of a client that presents itself as an autocratic director of the situation is an issue. Indeed, the emotional climate surrounding the organisational change and the building project can induce a compulsive drive in the client as a successful change agent. This elevates the client's power position to one of authority in the process, which may contradict the industry's recommendations. This power compulsion may manifest itself in the client seeking more from the project,

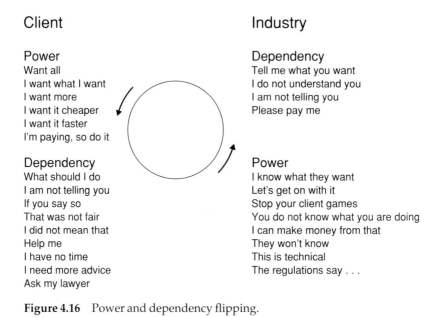

Figure 4.16   Power and dependency flipping.

industry is as a rescuer of the client. During the project the client sees itself as the victim and the industry as the persecutor. In the end, however, the client becomes a persecutor and the industry sees itself as a victim, and so it passes the blame among itself. This can be seen as flipping between power and dependency positions, as shown in Figure 4.16.

The client is often in a dependency situation. It assumes that it understands its own strategic planning and that it simply needs to purchase what it wants, namely a building. We suggest that many clients are complicit in this mistaken approach to building because of the emotion of building (see Section 4.5). The client may, at best, see a distance between the building and its organisational change, but more difficultly the client often does not see that it is undergoing organisational change at all. The scale of building means that it is generally a strategic management change that is occurring, which affects the whole organisation.

What of the expert client? A useful distinction that is made of clients is how often the client has built before, so that it understands the process and its responsibilities in it. At one end we have the client that could do it itself, with the role of the construction industry being simply to give labour resources, i.e. outsourced. Hidden in this, however, is the shedding of risk to the industry or a desire to ensure that if anything goes wrong, the client can muddy the waters enough for it to escape paying for the risk. To some extent, an expert client works both ways; it is good, as the client is familiar with the process and can engage to help itself and to allow the process to happen; on the other hand, the client can use its understanding to shed risk and catch the industry out in the complexity of the change process. The latter has hidden agendas, and its covert operation requires understanding in the cultural, political and psychological arenas in order to manage the client before it manages the industry.

Expert clients may appear to be rational in their decision-making, but this hides (pushes into the covert) the problems of culture, politics and psychology. Even for experts, building is an emotional process of change, involving high stakes in the investment in the building and also in the changes induced in the organisation. We only assume that the client's experience means that it has learned to deal with this.

And what of the inexperienced client? Maybe there is no such thing as a completely inexperienced client in the sense that it does not come in without any knowledge. Knowledge of the industry is built up from reports and others' experiences, which can induce a normalised response to the process. It is likely that this will be negative, as the image of the industry is negative and people tend to tell only negative stories of their experiences. Client innocence probably means that the client is unfamiliar with the processes and emotional ups and downs of the process; the client may not have the mechanisms for dealing with this. Added to this is the fact that building requires organisational development in the client – it is unlikely that the inexperienced client has the mechanisms to deal with its own organisational change, let alone engage with the problems of building.

Contract in client–industry engagement

The position of contract in client–industry relationships is problematic not because of the detailed content of contracts but because of the way in which contract mediates relationships. Contract is often presented as an enabler of relationships, placing obligations, rights and risk in a consistent, clear and agreed way (Murdoch and Hughes, 2000). In the logical rational route, this is the case and dictates that it is worthwhile taking the effort to make the contract as correct and as complete as possible, with suitable protections to enable ends to be created from means. This aspect of protection is important, in that contract provides a structured mechanism and a decision-making environment (including the Construction and Technology Court) when the means do not deliver the ends. There is a difficulty, however, even in the rational route, because contracts can never be complete or matter of fact, as the interpretation of its aspects is always indexical (Clegg, 1992). This incompleteness is not handled well by contract.

Contract also interacts with the unknown non-rational route. In this respect, contract can be an enabler or a weapon in the power and dependency relationship explained above. The aspirational and emotional aspects of organisational and building change are not embodied in the contract. Indeed, the incompleteness and the unformed unknowns are seen as errors by the rational logic of contract, which then initiates the protection mechanisms of contract to sort out issues. This can work, and modern dispute-resolution techniques such as adjudication and mediation can bring together the parties amiably in order to face the realities of the world. However, the aspirational and emotional aspects can intervene, and the loss of aspired ends or the annoyance at a means can create the desire to not resolve but to get one's aspirational dues. Contract then becomes a weapon for the parties to battle with, and the legal system is organised to undertake this. The fact is that contract is not neutral in the mediation of relationships, as contracts are structured to favour the powerful party, which offers the form for the less powerful party to accept. This orchestrates the battle, creating defensiveness by the least powerful party and a wielding of the weapon of structured justice by the powerful party. The legal system is also not neutral, as it works on the words of the contract rather than on the espoused intentions of fairness. Indeed, the system also preferences best argument, which can be bought through greater quantity of legal attention or superior legal expertise.

Thus, contract is sited in the rational route of means to ends, but it actually disturbs the non-rational route. This contradiction is endemic and introduces some of the negative effects of contract in managing the client–industry engagement. Clients that require apparent rationality in all matters have to face the incompleteness of contracts in defining and structuring all areas. This can involve attempting to make stronger and more complete contracts, which can ultimately make matters worse. More recent attempts at better relationship contracting through partnering can help by essentially removing the contract from the mediation of the engagement. There is still a latent problem if the plan gap or aspirational gap is too large and the contract is invoked and the problematic issues reappear rapidly.

4.8   WHAT IS TO BE DONE?

As we look at the client, we see a complex amalgam that may not be understandable – even the client may not understand it. The amalgam may be kaleidoscopic; as we turn, it changes and what we thought was the client has now changed. In such a world, we seek out fixed structure to which we want to relate. Maybe in doing this the industry makes a mistake, because it is the structure that the industry wants to relate to rather than a meaningful but complex representation of the client. This mistake causes a dislocation and is emotional. We need to understand clients better by undertaking a detailed study of their complexity; we will do this for six client sectors using the model of engagement.

REFERENCES

Albrow, M. (1997) *Do Organisations Have Feelings?* London: Routledge.

Alderfer, C.P. (1979) Consulting to underbounded systems. In *Advances in Experimental Social Process*, Vol. 2, eds C.P. Alderfer and C. Cooper. New York: John Wiley & Sons.

Argyris (1993) *Knowledge for Action: A Guide to Overcoming Barriers to Organizational Change.* San Francisco, CA: Jossey-Bass.

Bannister, D. and Fransella, F. (1971) *Inquiring Man: The Psychology of Personal Constructs.* London: Routledge.

Barrett, P. and Stanley, C. (1999) *Better Construction Briefing.* Oxford: Blackwell Science.

Bion, W.R. (1961) *Experiences in Groups and Other Papers.* London: Tavistock.

Boddy, D. and Buchanan, D. (1992) *Take the Lead: Interpersonal Skills for Project Managers.* New York: Prentice Hall.

Boyd, D. and Wild, A. (1999) Construction projects as organisation development. Presented at the 15th Annual Conference of ARCOM, Liverpool John Moores University, 15–17 September 1999.

Cardonna, F. (1999) The team as a sponge. In *Group Relations, Management and Organization*, eds R. French and R. Vince. New York: Oxford University Press.

Casti, J.L. (1993) *Searching for Certainty: What Science Can Know About the Future.* Grand Rapids, MI: Abacus.

Cherns, A. and Bryant, D. (1984) Studying the client's role in construction management. *Construction Management Economics*, **2**, 177–184.

Clark, P. (2000) *Organisations in Action: Competition Between Contexts.* London: Routledge.

Clegg, S.R. (1992) Contracts cause conflicts. In *Construction Conflict: Management and Resolution*, eds P. Fenn and R. Gameson. London: Spon.

Crichton, C. (1966). *Interdependence and Uncertainty: A Study of the Building Industry.* London: Tavistock Publications.

Encyclopedia of Marxism (2005) www.marxists.org/glossary/index.htm. Accessed 7 November 2005.

Fineman, S. (2003) *Understanding Emotions at Work*. London: Sage.

Flood, R. and Carson, E. (1993) *Dealing with Complexity: Introduction to the Theory and Application of Systems Science*. New York: Plenum Press.

Green, S.D. (1994) Beyond value engineering: SMART value management for building projects. *International Journal of Project Management*, **12**(1), 49–56.

Groak, S. (1992) *The Idea of Building*. London: Spon.

Hampden-Turner, C. (1984) *Gentlemen and Tradesmen*. London: Routledge and Kegan Paul.

Hargreaves-Heap, S.P. and Varoufaki, Y. (2004) *Game Theory: A Critical Introduction*, 2nd edn. London: Routledge.

Hutton, G. and Devonald, A. (1973) *Value in Building*. London: Applied Science Publishers.

Ibarra, H. (1993) Network centrality, power, and innovation involvement: determinants of technical and administrative roles. *Academy of Management Journal*, **36**(3), 471–501.

Iles, V. and Sutherland, K. (2001) *Managing Change in the NHS*. London: National Co-ordinating Centre for NHS Service Delivery and Organisation.

Karpman, S. (1968) Fairy tales and script drama analysis. *Transactional Analysis Bulletin*, **7**(26), 39–43.

Kelly, J., Morledge, R. and Wilkinson, S. (2002) *Best Value in Construction*. Oxford: Blackwell.

Lawrence, W.G. (1999) A mind for business. In *Group Relations, Management and Organization*, eds R. French and R. Vince. New York: Oxford University Press.

Leaman, A. and Bordass, W. (1993) Building design, complexity, and manageability. *Facilities*, **11**, 16–27.

Lord, A., Padma, N. and Birchall, D. (1990) How project managers perceive their role and the contribution of top management to project decision making. Presented at the 10th World Congress on Project Management, Vienna, June 1990.

Lovell, R.J. (1993) Power and the project manager. *International Journal of Project Management*, **11**(2), 73–76.

Macdonald, K. (1995) *The Sociology of the Professions*. London: Sage.

Miceli, M. and Castelfranchi, C. (2002) The mind and the future: the (negative) power of expectations. *Theory and Psychology*, **12**(3), 335–366.

Morgan, G. (1997) *Images of Organizations*. London: Sage.

Murdoch, J. and Hughes, W. (2000) *Construction Contracts: Law and Management*. London: Spon.

Obholzer, A. (1999) Managing the unconscious at work. In *Group Relations, Management and Organization*, eds R. French and R. Vince. New York: Oxford University Press.

Rokeach, M.J. (1973) *The Nature of Human Values*. New York: Free Press.

Rush, R. (1991) *The Buildings Systems Integration Handbook*. Oxford: Butterworth-Heinemann.

Saxon, R. (2006) *Be Valuable: A Guide to Creating Value.* London: Constructing Excellence.

Stacey, R. (1996) *Strategic Management and Organisational Dynamics.* London: Pitman.

Stringer, J. (1967) Operational research for multi-organisations. *Operational Research Quarterly,* **18**: 105–120.

Talbot, C. (2003) How the public sector got its contradictions: the tale of the paradoxical primate. *Human Nature Review,* **3**, 183–195.

Tannenbaum, R. and Hanna, R. (1985) Holding on, letting go and moving on: a neglected perspective on change. In *Human Systems Development,* eds R. Tannenbaum, N. Margulies and F. Massarik. San Francisco, CA: Jossey Bass.

Thiry, M. (2001) Sensemaking in value management practice. *International Journal of Project Management,* **19**, 71–77.

Thompson, J.D. (1967) *Organisations in Action.* Maidenhead: McGraw-Hill.

Walker, A. (2002) *Project Management in Construction,* 4th edn. Oxford: Blackwell.

Weick, K. (1995) *Sensemaking in Organisations.* London: Sage.

Winch, G.M. (2002) *Managing Construction Projects.* Oxford: Blackwell.

5 Property Developers as Clients

5.1 INTRODUCTION

Of all the sectors, property development is the most diverse. Each development client is almost unique in its approach to business. The developer's role is as an intermediary between the property, business and finance sectors. Essentially, property developers do not use the property for occupation but as an asset to be sold or leased. To do this, they need to use their own, or other people's, money, in order to acquire, design, construct, manage and market this asset. This sector has strong connections to both the finance industry, which funds it, and to the law, which protects it and helps it to operate. We restrict our attention here to non-residential property development. Although many of the larger developers mentioned in this chapter have a significant portfolio of residential property, they do not operate as residential landlords. Overall, it was estimated for the end of 2003 that there was about £611bn of property stock in Britain, of which £265bn (43%) was investment (IPF, 2003). Of this investment property, about £124bn was in retail, £100bn in offices, £30bn in industrial, and £11bn in hotels, pubs, leisure and services (IPF, 2003). This supports a large property ownership and development industry, which generates not only development work for the construction industry but also significant maintenance work. Much of the investment ownership is held by financial institutions such as insurance companies and pension funds plus a growing number of overseas investors.

The history of property development over the past 400 years started with large landowners making money from agricultural tenants and subsequently mineral resources (Rosenheim, 1998). Their rights to ownership and exploitation were set down in law as society moved from a feudal to a capitalist structure in the seventeenth and eighteenth centuries, although ownership often started as a gift from the monarch for services. The move to being able to obtain rent from property and to own and trade in buildings was generated at this point. That a building itself is an asset requires someone else to have a need for and pay for that building. Buildings always had a storage function, and therefore warehousing was one of the first commercial building types to become marketable, either as a result of agricultural harvest or from sea trading. Indeed, wealthier merchants become a significant property-owning class. It is the advent of manufacturing, professional and now retail uses, however, that has driven property development to its significant status. Businesses choose to rent buildings because they cannot afford to acquire or build in these locations. The need for commercial businesses to be sited together in cities and to have access to transport makes location particularly important. Property built on a larger scale can accommodate several organisations and thus can be cheaper for a particular location. However, owning buildings as assets requires the management and maintenance of those buildings, and this generated new professions of estate management, land surveying and building surveying.

There is no recognised breakdown of types of developers, and so pragmatically we identify three groupings: property companies, traders and smaller property organisations. The distinctions between these are scale, financing, time horizon and personnel. The largest developers are property companies, which use mostly stock-market long-term capital finance, retain the properties in a portfolio over time and are managed by a professional corporate team employed by the company. Property companies include many big players, with Land Securities, MEPC, Hammerson, Slough Estates and British Land being the top five. In 2005, Land Securities had approximately an £8bn portfolio, which was some 40% larger than Hammerson's, at £4.6bn, and included a substantive share of the property-outsourcing market (Land Securities, 2004). MEPC is wholly owned by Leconport Estates, a company owned by clients of Hermes Investment Management Limited, primarily the BT pension scheme. This company used to be number 2, but it sold off a considerable asset in order to concentrate on its business space sector. In addition, international players such as ProLogis are investing heavily (in distribution in this case), which is altering the market and, more importantly, changing some of the draconian lease conditions that have been famous in the sector. These clients are all corporate, in the sense they are controlled by shareholders. Although major companies themselves, retailers and manufacturers are clients of these developers, so too are the government and other public authorities. In a sense, private finance initiative (PFI) and other public–private partnership (PPP) schemes are a wider form of property development, often in specialist markets (prisons) and often involving maintenance and operational services. The government and companies want property off their books because of changes in accounting practice that favour converting debt to a recurrent cost. The Crown, the Church and several large families' estates (often held in trust) operate as corporate developers of both rural and urban land. Developers such as Grosvenor, although private and effectively owned by the Duke of Westminster, are as large as public property companies managing assets of £7.7bn in 2004. Many private estates are secretive about their holdings and have agents acting for them.

The property trader developers are small, use mainly short-term bank and institution loans, do not retain the building, and are managed by small professional teams, which often have a stake in the development. Property traders find land opportunities, create a development plan, sell the idea to a funder, manage the development (including planning, design and construction), let the development and then dispose of the development. They make their money from taking on the risk of the development and letting processes that funders are unwilling or unable to do. The properties are sold with tenants into the property portfolios of others.

Finally, small developers can be individuals, families or small investment groups. They use mainly their own finance, often obtained from other enterprises, e.g. legal or accountancy practice. They may or may not retain the property in a portfolio, but the portfolio is not substantive enough to be a risk-mitigating device. They may have no professional training but can be experienced business-people. The developments are ways of making money for individuals or for providing actions that hedge the core business activities.

5.2 BUSINESS ENVIRONMENT OF PROPERTY DEVELOPMENT

This chapter deals only with property companies and trader developers, as small developers have so many differences that they are difficult to encapsulate. However, we will also consider a specialist trader

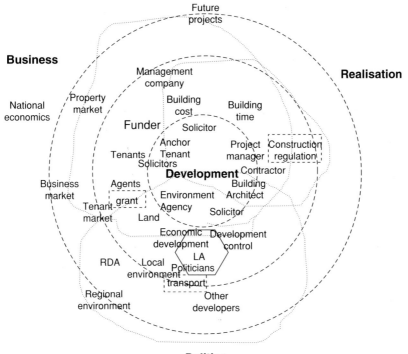

LA, local authority; RDA, regional development agencies.

Figure 5.1   Map of property developer's environment.

developer that works on mixed-use regeneration schemes. These are
private-sector organisations and so can be modelled by business-system
hierarchy theory involving strategic, tactical and operational levels
(see Section 3.3). The map in Figure 5.1 demonstrates the complexity
of the actors and issues that inhabit the property development world.
We have isolated the strategy around the business plan, the tactics
around managing the politics, and the operations around realising
the building, which includes construction. The world will be slightly
different for the property company and a trader developer associated
particularly with their relationship with finance; however, with regard
to a building realisation, there are fewer differences.

   Looking back on successful developments, it may appear obvious
that they would have been successful. However, there would have
been considerable doubt about the success at an early stage when the
development was only an idea. For the idea to be successful, a number

of factors need to come together to make it possible to even go ahead. The way in which these factors come together determines how the development is composed. Strategically, these are (i) making it work in business terms, (ii) making it work in political terms, and (iii) making it work as a building. The problem is that these aspects interact, and changing one affects another. A successful development means that a solution that satisfices in each area has to be found and that these act together to form a whole. To some extent, they need to be considered together; most developments, however, are driven by the business plan, which is the motivator. Many issues cut across these factors, including land purchase, planning permission, tenant market and building cost. Success means meeting the demands of each in some way and then moving on, even if they are not ideal, as delays may lose the impetus for the development. Each of the actors and issues needs to be managed, and the relationships and importance change over time.

Business of property

If property development was simply business trading, then it would be easy to understand, namely that the money return had to be greater than the money spent. Money is spent not only on the land and the construction but also on the interest on the finance. There are many complicated financing arrangements that funders of property development enter into associated with the degree of risk and control that they desire. Regardless of whether the funding is internal or from a funder, the development is expected to make a return on investment that is greater than the prevailing loan rate. The money returns through using the property as an asset, which involves two options of making money: there is a capital increase in the value of the property and there is an income from letting the property to others. The first of these is realised only when a property is sold, but it is declared as part of the assets of a company benchmarked by an annual valuation, which can influence share price. Historically in Britain, the value of property has risen above the rate of inflation because of the shortage of land and stringent planning regime, and this has made property a very secure and lucrative form of investment, particularly against more volatile stock-market investments (IPF, 2005). In addition, the security of financial arrangements, the fluidity of a property market, and a good and defendable structure of ownership make Britain a favourable property business environment (IPF, 2005). This has encouraged a growing overseas investment in UK property.

Maintaining a property portfolio for rental is still one of the best ways of making money, and so banks, assurance companies and pension funds as well as property companies are willing to make significant property investments. Property owners may use this income to fund their next development, or they may be completely detached, possibly never being at or even knowing where their investment is. Thus, as well as a market for property sales, there is a separate market for property rentals. There are some connections between rental values and the sales values of property, although rental is more volatile, depending on general business needs and business success. Demand within a market is a function of price, availability, substitutes and tastes. What is being rented out is not only space but also the location for business, either for customers or for labour market, plus a quality of appearance. There are blue-chip clients who are willing to pay higher rents and accept longer leases and thus make a development more attractive for selling on. The average length of leases has reduced and is now about eight years for offices and 12 years for retail (IPF, 2005). The building user market is differentiated into sectors – offices, retail, leisure, industrial, residential and distribution – and this is reflected in planning use classes (Cullingworth & Nadin, 2002). Offices are the tradition for property companies because of their association with blue-chip clients and greater certainty of return. Since the 1990s, retail has seen a rising market starting in out-of-town centres but is now seen as part of the regeneration of towns; its market is tied to consumer spending. Attracting a major store to a development encourages others to rent space, due to the increase in footfall as more consumers are attracted. Mixed use, involving offices, retail and residential, is currently in vogue, as it matches planning needs and offsets risk to the market; however, it requires space to be used differently and makes development more complex. Logistics and warehousing are a rapidly developing sector because of new approaches to retailing, with buildings sited near transport interchanges but feeding regional outlets.

As well as developers spending money on goods and services for the development, they also have to pay taxes, which can be quite considerable. There are taxes involved in development (e.g. VAT), building purchase (e.g. stamp duty and land tax, which can be up to 4% of price), building ownership (e.g. commercial rates, death duties), building sales (e.g. capital gains up to 40% of profit if an individual) and development business (e.g. corporation, which was up to 30% of profit). Some taxes, however, can be offset by capital allowances available on plant and machinery, buildings – including converting space above commercial premises to flats for renting – and research and development. It is possible and worthwhile to design the development or purchase the development in such a way as to reduce exposure to taxes (IPF, 2005). In addition, there is a tax on occupation, i.e. rates, which are calculated from rental value. Although rates are paid by the tenant, they form part of an overall occupation cost, which a developer can influence in a number of ways. The occupation cost includes an annual service charge, from which some developers may choose to make a profit. Thus, there are opportunities to create deals for tenants involving lower rental charges but higher service charges, whereby the lower rental can yield lower rates on the property.

Grants are an additional potential source of funding. For example, urban development grants are discretionary grants used for promoting job creation, inward investment and environmental improvement, by developing vacant, derelict or underused land or buildings in priority areas. There may also be grants for remediation of contaminated land in certain areas. The issue for the developer is that there are obligations attached to receiving grants concerned with jobs, provision of particular space, rental conditions or conservation, which may restrict other aspects of the development. These restrictions may mean that the use of grant money, even if it is available, may not be worthwhile.

With regard to the land, a developer may desire to purchase more in order to have a better site, whether for scale, shape or access. Land purchases around a potential development site are notoriously difficult, as the developer is in a weaker negotiating position when the development is made public: the surrounding landowners try to exploit this in order to achieve a higher price. Land assembly involves bringing together plots of land to make a viable site, in terms of scale and form, which requires purchase from multiple owners or land swaps, requiring much negotiation, pressure, legal involvement, money and time. Easy sites are, therefore, much sought after; however, these tend to be in less desirable areas or are extremely expensive.

Political and regulation

Property development is a highly political area of business. It impacts on people other than those involved directly and on the environment.

In addition, it is identified with social power, which can be used to make high financial returns. Thus, it has become highly regulated and monitored, which interferes with the business plan. Politics can be regarded as a system of handling conflicts of values with regard to needs, actions and rewards. These needs and rewards are different for different stakeholders, and political negotiation attempts to secure reward for, or prevent damage to, particular stakeholders. A social stakeholder is identified (the benefit to society), which has a wider and longer-term objective than the development itself. This social stakeholder is represented by regulations and political factions. Thus, a development can be obliged by regulation to maintain a landscape or quality of environment. In some quarters, property ownership and property development are viewed as an exploitation of society and a maintenance of class and power positions (Cahill, 2002). Thus, property development has the potential to cause great emotion and conflict within a political domain.

Property supports all aspects of other businesses involved in the local and regional economy (BPF, 2004). Thus, it is the precursor to much infrastructure and individual business development while also providing significant opportunities for employment. The timely and efficient production of buildings and their operation can be a major supporter of business and services and, hence, the local community. Thus, the creative development of property, within its financial, legal and social constraints, establishes both a physical and an economic impetus to regeneration. The bigger the scheme, the more likely that politics will be involved, as regional economics are affected by developments. This is particularly the case in areas of higher unemployment, as anything that reduces this is seen as critical. Political support for a scheme ensures a smoother passage and the overcoming of problems, which are bound to arise. Political opposition means that many steps may require a fight, and problems may be seen as opportunities to scupper the scheme.

The use of land is also a political issue because of either detracting from or enhancing the surrounding value. Development and business are seen in some quarters as detracting from the public good. The main vehicle for sorting this out is the planning system. Planning is the major set of regulations constraining development through limiting the extent and the aesthetic of what can be constructed (Cullingworth & Nadin, 2002). Thus, although a scheme can 'stack up' financially, it will not be possible to go ahead unless it aligns with the local planning position. All developers therefore compose not only a business case for their schemes but also a case demonstrating positive local impact. This can become the centre piece of the developer's vision, particularly if it is a regeneration scheme. This also involves considerable lobbying, public relations and media battles with opposition groups. The degree of organisation and sophistication of these opposition groups affects a development's strategy, which must involve encouraging those that support the development. Many developers employ public relations consultants and political consultants to present and manage their cases. This battle for support takes place early on, at the same time as the financial case and the land purchase are happening. This must all be right in order for the scheme to go ahead, and this can require considerable time, energy and inventiveness.

Planning can also require a degree of investment in the local infrastructure, through Section 106 agreements (Cullingworth & Nadin, 2002). Many developers view this as a tax, and there have been several suggestions to formalise it into a development gain tax. These infrastructure obligations might include new roads, junctions, community facilities, art and social housing. The degree of this can intervene in the financial plan of a development, although fighting local planning decisions (going to appeal) can be expensive and time-consuming. Other regulations, which are becoming of great importance, surround environmental issues. The Environment Agency can impose conditions on a development, with regard to soil remediation, maintenance of water resources, including aquifers, flora and fauna, and air quality. As well as this, there are responsibilities for archaeological remains, which may be found during the development. These external issues, although involving regulatory bodies, are part of the unknowns of the process of development.

Development realisation

The process of development realisation involves a trade-off between business, politics and user needs. Thus, increasing the building cost means that interest payment will be more, but this may be worthwhile if having a higher-quality building makes it more lettable and easier to obtain planning permission, or makes the construction faster and so allows earlier sale or letting. Higher costs per square foot, however, mean that rents will need to be higher, which makes the building less attractive in the marketplace or reduces profits. By concentrating on certain use classes, then, a higher economic impact could be demonstrated in an area; thus, planning permission might be easier to obtain and the political will might back the marketing of the development. However, these uses may not be possible to let at a profitable rate. These trade-offs can be, and often are, regarded as opportunities or constraints that need to be fought or circumvented. It is this bravado that is often characteristic of trader developers and that sees them offer creative solutions in order to overcome the constraints of regulation. This process of trade-offs can infect the scheme for the whole of the project, both emotionally and financially, particularly as the building nears completion. This emotion infects all parties involved: the political stakeholders, in case they do not get their public relations benefits; the funder, in case their return is less; the developer, in case they are left with an un-let building; and the constructors, in case they do not get paid.

5.3   FINANCE AND RISK OF PROJECTS

As indicated earlier, business planning is affected heavily by the way in which a development is financed; this determines what business success means and how this can be achieved. This also establishes the project risk and the consequences of events not occurring as planned. The financial arrangements set a time framework under which results have to happen and a pressure for certain achievements in time. For example, the time framework of the financing becomes the time framework of the development. Property development takes place over an extended time in comparison with other financial investments, and there are many external and internal factors of development that make the investment volatile and riskier in the short term. The reporting and accountability associated with the financial package can be onerous and invasive. Risk of not achieving a successful development is apportioned by the financing, to be held by different players, who try to pass this on. The funder passes risk to the developer, who passes it on to the design and construction organisations, who pass it on to subcontractors. The way in which different players perceive risk (Kloman, 2000) as something to be abhorred or welcomed categorises the different players in the sector and determines the way in which they think and act.

Financing involves complex packages of arrangements, so that the funder is protected (Isaac, 1994). The differences in the form of finance have a great effect on the way in which decisions are made and the way in which stakeholders behave in the face of events.

The basic form of finance is *debt finance*, which means that money is loaned to be paid back, regardless of whether the development is successful. Interest rates are declared (fixed or variable against base rate), and the time over which the transaction is operational and the repayment dates are fixed. Riskier developments and less experienced developers will be charged a higher interest rate. Of course, there is a risk for the lender of the debt not being repaid, and so a *security* is required, for example that the lender has a legal charge in case of default to sell the property to recover their money. If only the development is involved in the legal charge, then the financing is *non-recourse*; however, the lender can reduce their risk by requiring *recourse* to the wider assets of the borrower.

On the other hand, *equity finance* involves the lender taking a share in the risk and profit of a development. The returns on equity are more variable and insecure but can lead to greater returns. Equity financiers normally expect a say in decisions because of the extra risk involved. They may not be property people; thus, their decisions may not take into account the complications of development and construction but may be more concerned with their financial needs. This may constrain the development objectives and the methods of achieving them.

Funding of a project usually will utilise a mixture of debt and equity funding. Debt funders will finance up to 70% of a project and so expect a developer to have a percentage of equity in any project in order to reduce the debt funder's risk. If the developer cannot raise sufficient equity capital, then they can access *mezzanine finance* (possibly up to about 15%), which is similar to debt funding but more expensive because of the greater risk; it also commands a return from outturn profits. The financial packages that a developer creates may change during the development and involve different rates of interest because of the different exposures to risk. They may require different finance for development and for the void period. Indeed, if the developer is using the sale of other property to finance a project, there may be gap funding between the need for the money and the sale of the property.

Joint ventures involve two or more organisations creating a new organisation through which they share the funding arrangements. They may bring different things to the project, such as land, equity, cash and management expertise. The company that is set up as the vehicle for the joint venture has an existence of its own and, however small, it requires its own management and reporting structure. Decisions are then made through the structure of the joint venture and so are constrained by its arrangements and the various stakeholding needs of its parties, some of which might be in conflict, either in outlook or in time. Newer forms of financial vehicles for larger developers, such as property unit trusts and real-estate investment trusts (REITs), require formalised corporate decision-making processes.

All developers use some form of development appraisal to support their business plan (Isaac, 1996). These will be in more detail as the development progresses towards being given the go-ahead. Development appraisal involves calculations, which start with valuing the completed development. If the development is for sale, as in the case of housing schemes, then the costs of undertaking the development, including land, construction, cost of finance, cost of marketing and cost of time to sell the properties, are subtracted from the sales returns. In developments that are to be operated as rental schemes by the developer or to be sold on as rental properties, then the calculations are more complicated, as they involve making estimates of returns over time. This takes historical data modified by local conditions and determines both an expected rental value and the market yield to calculate a gross development value. Again, from this are subtracted construction costs, consultants' fees, costs of finance, legal fees, letting fees, investment sales fees, advertising fees and land costs in order to determine the developer's profit. The calculation can be done in another way whereby a profit is fixed and a maximum cost of land that can be paid is determined. Such calculation can be undertaken in great detail, but all use a high degree of estimated figures. More sophisticated calculations are possible, which embody the degree of uncertainty of the estimated figures. This at the simplest involves a sensitivity analysis, in which the estimates are varied to ascertain the effect on the viability of the development. At its most complicated, it involves such techniques as Monte-Carlo simulation, which produces a statistical estimate of confidence in the expected figure.

5.4   BUSINESS CONSTITUTION: STRATEGY TO OPERATIONS

The way in which a property development business is structured and its strategy determine how it acts. In this environment, then, the change that is being sought is the delivery of greater rental value or of greater resale value. After the capital investment, the prime characteristic of this is time. In a classic trading way, it is possible to make money by buying and selling very quickly before anything has been done. There are, of course, transaction costs (mainly legal, but also duty payments) that limit the desirability of this. One aspect of development, however, is making money from the downgrading of risk; for example, it is easier to sell on a fully let building at a higher price than a building that is un-let. This takes time. Whether this is the design time, the construction time, the letting time or the sale time, the shorter the time can be, then the quicker there is a return on the investment. As external conditions change – financial or regulatory on one side and business and desirability on the other – then so do the economics of property development. These changes over time induce a degree of uncertainty into the process and a potential to have to change course as a result.

Property developers strive to achieve a higher rental or sales value by creating an enhanced facility or a better match between the facility and the needs of a particular tenant. If a building is pre-let to a tenant before construction, then a developer will be able to match the building better to the tenant's needs, but with a view to an alternative use if the tenant leaves. An enhanced facility might be produced through delivering aesthetic acclaim or fitting the building creatively into its location. The match may, to some extent, be due to marketing (i.e. telling people what is there) and to market awareness (i.e. building what people want); it is, however, also a sales function (i.e. convincing people that they need what is there) (Newman, 1997). It is a subtle developer skill to deliver what will be wanted by both tenants and buyers. The first is critical, but it must not be at the expense of deterring the latter, as the equity increase is most valuable.

Thus, strategy, tactics and operations have different meanings in property development, where the business of the client is the building. Of course, it is not only the physical realisation of the building that is important, but also the realisation of the building as a business, i.e. a successful building as a business is one that is let or sold on at a profit. This condition affects different types of developer in different ways.

Property company

For a property company, strategy lies in portfolio management. Indeed, property may not be the only vehicle for financial profit but may simply be one form of management of assets. Asset management has some confusion in the property world. From a financial investment perspective, assets can be seen as the set of all items that have value. Thus, stocks, bonds, contracts and leases are assets as much as buildings or machines. When property companies talk about assets, they are probably talking about the financial substance of their investment, which is somewhat abstract and may have a variety of forms. However, in building management terms, assets comprise the physical building and equipment. This problem of perception and language can cause communication difficulties, as there is a difference between managing physical things and abstract things.

Development is a small part of a property company's activity, and often buildings are added to a portfolio through buying a ready-let building. Portfolio management involves the selection, purchasing, performance appraising and disposal of assets in a set in order to meet wider objectives (MacGregor and Hoesli, 2000). Thus, it involves balancing opportunity and risk across the portfolio in order to achieve the maximum return and strategically selling or purchasing assets in order to undertake this. In property portfolio management, the balance may be achieved by having property in different markets (retail, office, logistics), different locations (city centre, motorway junction, London, New York, Shanghai), different exposures (joint venture, sole owner, bond-financed) and different lease conditions (short-term remaining, full repair and insurance, blue-chip tenants). As the assets are seen as abstract, then the portfolio performance is seen as financial, such that econometric calculations can provide an assessment of the position of the whole investment and this can be compared with investing only money. It is the portfolio as a whole that is of concern. It may mean that trading is required in order to optimise the portfolio, i.e. both buying and selling of abstract properties to balance or optimise risk and return to enable a secure future. This is based on actual performance or forecast predicted performance. Modern portfolio theory (MacGregor & Hoesli, 2000) produces an overall investment strategy that seeks to construct an optimal portfolio by considering the relationship between risk and return. This theory recommends that the risk of a particular stock should be looked at not on a stand-alone basis but rather in relation to how that particular stock's price varies in relation to the variation in price of the market portfolio. The theory goes on to state that given an investor's preferred level of risk, then a particular portfolio can be constructed that maximises expected return for that level of risk.

This desire to, and belief in their abilities to, rationally protect themselves from uncertainty requires property companies to be structured as functional bureaucracies with a high level of procedurised operation. Most property companies have a holding company, below which are the operating companies, including different country- and sector-based companies and other business, e.g. insurance. The holding company acts as the international brand owner, with the operating companies being almost independent. The holding company board has the chief executive officers (CEOs) of the operating companies and will have non-executive directors in order to ensure its propriety as a publicly listed company under new corporate governance regulations. The operating company board is much more functional but is still driven by finance, as it differentiates between major income and spending divisions; it is this that determines the constraints that a company sets itself under which to operate. In the example in Figure 5.2, there is a funding division that seeks resources and partners, and then different market-sector divisions, such as retail and commercial. Most property companies separate their London properties from the rest of the UK and then subdivide this into sectors, as this is such a specialist market. Development is a division on its own, as it is a major spender, although it will work across the other divisions. To support all this as a business, there is a corporate division, which may include legal, tax, research, procurement, human resources and corporate management.

Each business division carries out highly detailed evidence-based and quantitative assessments of its activities. The strategic questions to be answered are: How much should the company be exposed to property? How does the company maximise the long-term return but also maintain an annual dividend? What mix of properties will achieve this? Should higher-risk developments be undertaken, or should only buildings be purchased? There are other calculations that surround the corporate overhead; for example, a strategic decision such as concentrating on one sector (selling property that is not in this sector and purchasing property that is) allows a company to maintain a greater understanding of this market (i.e. reducing risk at a lower cost). This could mean changing the company structure and so induce resistance from divisions. Their case in opposition will be that this places dependency on one sector, and thus the volatility of the sector affects the operational profit (dividend). Such decisions on the composition of the portfolio therefore induce paradoxical decisions about particular developments at any particular time, all the way down the company, which may be due not to that development but to the wider portfolio and the market conditions at the time. There is the potential for all

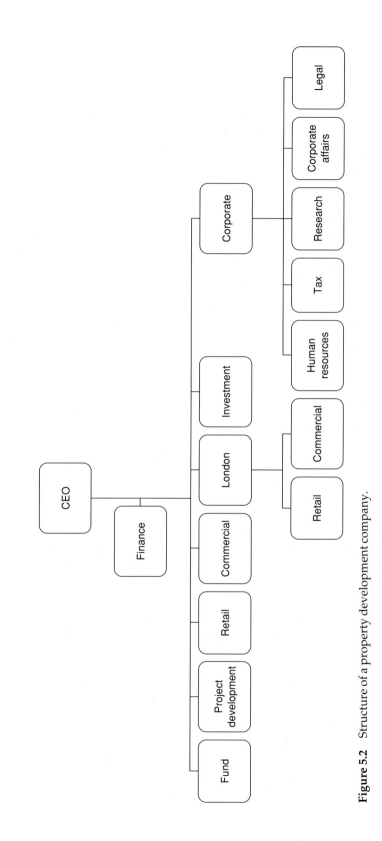

Figure 5.2 Structure of a property development company.

corporate organisations to have contradictions between strategy, tactics and operations. It is the way in which they manage these contradictions that determines how they operate.

Tactics for a property company are implemented through the different divisions that are set strategically within the company and concern the way of managing a particular section of the portfolio. This may be area-, building functional type-, tenant sector- or organisational function-based. The setting of rents in a particular sector and under particular market conditions, and the offering of incentives, can expose differences between the needs of the different divisions. The business for most of the divisions is not development but the finding of tenants, managing income and managing leases, with the strength of the covenant being paramount. There is little interest in the operational concerns of tenants themselves unless the tenant is large or provides an anchor, as then the tenant relates to the risk within the portfolio.

Property trader developer

In a trading company or mixed-use regeneration developer, strategy is concerned more with the next development. There may be a decision that the company will work only in certain locations or only with certain market-sectors (e.g. offices or retail). If a suitable development opportunity or piece of land arises, then the strategic decision will be whether to change the company in order to cope with more work. Riskier developments require concentrated attention in order to realise the opportunities, and thus there is a limit to the number that can be handled. Opportunities are sought, but at this strategic stage the realisation of them is full of uncertainty: they are only a vague idea or an intuition that something can be made of this. Such strategic decisions rely on people trusting the skills of trader developers to access finance and even to take an option on land, and this is all part of the business capital that they are exploiting.

They are very small dynamic organisations involving four to eight people, often employed by a pair of directors. The people tend to be multiskilled, but there may be a land buyer, a surveyor, a development project manager and two or three support staff. These staff are not sufficient to realise the development themselves, and so property traders need consultants to provide their organisational expertise, to mitigate risk and to provide credibility. They have a dilemma as to whether to employ cheaper consultants or to use larger, more costly consultancy practices, as their involvement adds credibility to the project. As property traders are small organisations, strategy is easily communicated and appreciated: the strategy for a development is merely a vision of success, which itself changes during the development and will involve the whole of the organisation.

Tactics are part of operations or strategy for the traders and mixed-use regeneration developers. There is no intermediate level of business, as the two or three developments that these companies are undertaking take up all their resources and are seldom interrelated. There are tactical decisions for implementing the strategy in a particular development, which is about developing courses of action against known and unknown eventualities and which requires compiling a detailed understanding of the local market, local political directions and funding sources. This relates to the individual trader developer's way of doing business.

Mixed-use regeneration developer

Mixed-use regeneration is a specialist form of trader development. The projects tend to be bigger and carry more risk and, as such, require a greater degree of attention. The regeneration developer is working more at the edge, and therefore their personality needs to cope with this. They are highly motivated people and expect the same of those people who work with them. They need to be more creative than normal trader developers, as their vision has to inspire in order to induce change in an area. They are experienced, but the conditions of regeneration and mixed use means that they cannot know everything. They are undertaking schemes that are larger than the substance of the company. They require a very able consultant team to inform them and help them to make decisions. They need people to tell them when they are wrong but also people who can accept that when they make a decision – even if it is off the wall – then they need to go for it. Using consultants from large practices and large contractors gives the project credibility and allows some risks to be passed on through procurement via warranties. It also defends the small developer from attack if things go wrong, by having strong and wide expert advice and a degree of substance from the wider organisation in order to stop it being pushed around.

Success requires an intimate understanding of the market, plus seeing and taking opportunities from this as they arise. Regeneration land is cheaper than better located land, and so there is the opportunity to use some of the potential value rise to invest in buildings. In addition, this also provides some resources to invest in the surrounding infrastructure. This makes the potential regeneration more acceptable politically, as the area is being upgraded, and it also presents a positive image to the regulatory authorities. This positive position can enable a more flexible planning consent, which allows opportunities to be taken as they emerge and also can ensure a substantive amount of political support, which assists with marketing. Restriction on the quantities of uses may be applied by planning, and this factor may be a constraint. A mix of uses allows a degree of diversification and thus risk-mitigation as the marketplace changes. The actual mix of mixed use can determine the financial viability and the sustainability of the development. Although returns can be higher or occur sooner with some uses, the overall development needs to work as a whole in order to maintain the vision. There is a desire to deal with whoever comes along to create a package of opportunity. The dilemma, however, is whether to seize opportunities to let the space or to maintain the vision.

Against this desire to exploit opportunities as they arise is the funder's need for a solid business plan and controlled targets. As regeneration may not yield to such bureaucratic control, it may be necessary for the developer to take a higher risk in order to retain control. Thus, the developer may choose a substantively debt-funded project so that decisions do not need to be referred to the funder. This also reduces the documentation for accountability, which can cause a substantive overhead on the development.

Regeneration developers need to be very hands-on, which means they must be intimately aware of all aspects, from finance to site progress. They have small teams that work together closely and communicate quickly and decisively. There is often a degree of serious fun that reflects the high-risk character of projects.

5.5   EXPERIENCE OF BUILDING: FROM UNKNOWNS AND CONTRADICTIONS TO MEANS AND ENDS

It is now possible to identify the unknowns and contradictions in the way that developers are constituted and operate that affect the way in which they engage with the construction industry. In other sectors whose business is not development, there is always a gap between the business and the development. In property development, this gap is less clear. The business unknowns and contradictions directly influence the means and ends of the development. The business change is the development, and so this building and the business must be managed together.

Property companies

Property companies are large corporate organisations structured into functional divisions. Satisfying this diverse organisation is paramount but problematic. In many ways, the building is an incidental or tangential concern to the business, which is focused on investment. Apart from trophy buildings, which tend to be in London and other major centres, the significance of the individual building depends on the current need for expanding (or diversifying) the portfolio or consolidating it in a calculated manner. This strategic contradiction induces a distance between the building developing as concrete substance, with the problems of abstract financial realisation. Figure 5.3 shows the means and ends of property companies.

The company's knowledge and expertise are extensive, but these are held within each division. The most valued company knowledge centres on financial management, portfolio management, development appraisal, risk analysis, structuring of leases and relationships with corporate clients. There is detailed construction management knowledge and experience in divisions; however, development accounts only for a small part of the business (maybe 15%) and so will not command a high authority. This authority is enhanced by the public-relations function of new development, as it demonstrates the company's investment in renewal and economic infrastructure.

Most business processes surround the achievement and maintenance of stability. In corporate organisations, action must be planned and justified within a rational and quantitative framework. They are pillars of the establishment and need to maintain not only economic success but also a business image with their investors, their clients and the public. They will have implemented corporate governance rules, have a corporate social responsibility policy, and be concerned to have an environmental and sustainability position in all developments. Each of the functional divisions has a set of procedures for undertaking its function and a set of calculations to enable it to make and justify its

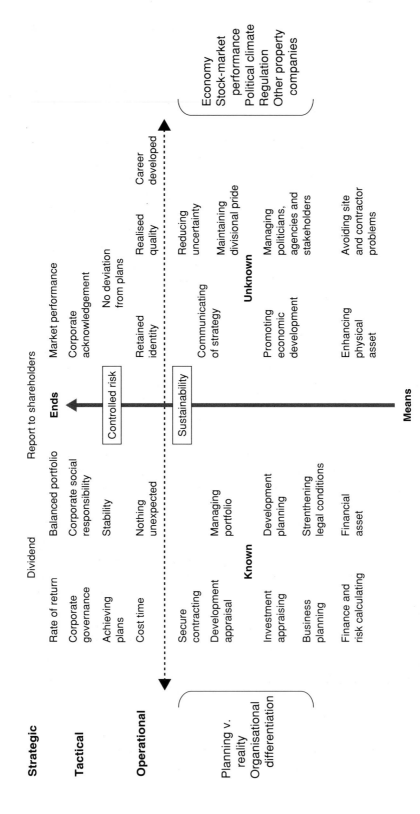

Figure 5.3 Means and ends of a property company.

decisions. This is administrative management and as such requires the proper process for action and accountability of action. Decision-making involves deliberating over the positions of each functional division within the hierarchical committee structure, and there will be (or will be a search for) an overarching quantitative model, which will make the final decision. Thus, planning and decision-making are slow and deliberate, and with little opportunity for flexibility. There is a need to be seen to be in control and for taking lower-risk approaches. Calculations and rules dominate in property companies, but there are difficulties in making decisions that are marginal and require judgement. The development appraisal and more complicated quantitative assessment methods are important in decision-making. The more sophisticated the appraisal, the more accurate it is believed to be. These expose a conflict in attitude within divisions and among individuals between theory, intuition and the developing reality. Any higher-risk decisions are passed around so that no functional division takes responsibility, or there is a desire to transfer the decision to consultants in order to protect the internal system.

Communication across divisions is difficult in all corporate organisations and causes incomplete and delayed decisions. The development division has to be more aligned to the development goals but may be less aligned with each division's desires for the development. The relationship with tenants and meeting their needs can be displaced in the company, but the pressure from the tenants comes directly into the project. The property company is protected against tenants through their insistence for strong leases, which can induce disagreements between pre-let tenants, agents and the company, especially if the tenants do not get what they expect. The construction industry is then caught in the middle of a commercial disagreement. In an opposite way, but from the same cause, portfolio and investment issues can interfere with the development. The success of the development is reviewed against stock-market conditions, which, if volatile, can induce nervousness in investment. In addition, the reporting cycle of public companies needs to demonstrate positive performance in reports and needs to justify action to shareholders. The stock-market changes may induce a requirement for a short-term return against the longer-term nature of the project. The company is less flexible as these reporting times loom.

There is a desire to have great certainty, so that as problems occur the emotional climate becomes negative if the system is disturbed. This climate requires extensive reporting that critically evaluates action and management to return the development to stability and control. If it is a company mistake, then they have great resources to defend their corporate position. There will be considerable embarrassment in the division that this disturbance took place in and so there is a potential for covering backs and passing the buck. Anticipation of this reaction, as a problem emerges, may paralyse the organisation. The pedigree and financial strength of the company will always overcome problems. They will use institutional power but are willing to pay to change meaning that covers backs. Cost may be the least important of the issues, although it may be the instigator of issues; more important is adequate and acceptable explanation, good reporting and proof that the correct control procedures have been undertaken.

There is a corporate career ladder. Promotion comes from qualifications and from successful management of company affairs. There is a degree of movement of personnel around the property companies, and senior posts are filled from other property companies and consultancies. Many senior managers will have MBA degrees from prestigious universities and involvement at a senior level in professional property activities. People are moulded by the corporate situation and are under pressure to meet targets and to advance their careers. New development provides an opportunity for individual career success in the divisions. This pressure feeds through into creating volatility in attitudes to events in the construction project as they occur.

The property company experiences buildings as concrete and unmovable but sees them as a commodity that is flexible and transient. This paradox is not managed but presents itself in many decisions about the building. Building needs to represent the image of the company, but it is also a fluid asset with only abstract meaning. Rational decision-making in committee tends to lower the novelty of a development, as divisional conflicts are met by the lowest common denominator. Buildings are delivered to institutional specifications. In a similar way, corporate organisations require a certainty of planning, with no surprises. They work in a world of understated disputes, where the system is allowed to dominate. This causes a degree of hidden issues that lie latent at the boundaries between divisions. The gap between the organisation and the details of a building project can be large. This gap can be filled by the construction industry. Thus, partnering as a means of procurement allows the industry to act with the property company. Partnering reduces the risk of uncertainty at the abstract corporate level by containing the uncertainty, at least psychologically, within the relationship with the partners. Problems then can be worked through without inducing corporate conflict or emotion. Consultants are then viewed as risk-mitigating partners.

Trader developers

Trader developers work for funders by taking on the risks that funders will not take on, by doing the onerous tasks and by pushing the regulatory boundaries that funders are unable to push. Figure 5.4 shows the means and ends of trader developers.

Trader developers are constrained by having to report to and get decisions agreed by the funder. Funders are risk-averse and think differently from the developer. The package of financial arrangements with the funder determines success criteria and timescales, which the developer translates into the realisation of the development. Often, the developer is forward-funded, in the sense that the funder agrees to purchase the development once it is complete and let, while charging the developer interest on the finance. These arrangements produce a profit erosion for the developer; as time goes on, the profit can slip, which induces an anxiety in the development.

In the current cash-rich, low-interest world, institutional funders are in competition with each other to offer finance to safe investments. This competition may force them to fund developments that are not within their usual domain and are riskier. Traders take this opportunity to obtain funding at the lowest rate possible. If a development has a more favourable appraisal, then they are more likely to be able to negotiate better rates from funders. The development appraisal and more complicated quantitative assessment methods are not as accurate as presented. Funders require data that can be justified, but many of which have to be estimated. It is possible for developments to be moved from a marginal decision to a clear positive decision by adjusting estimated values in the appraisal. Small adjustments, say to market yield or rental value, can significantly alter the favourability of the development. This means that the judgement to go ahead can be challenged at a later date. Big lenders are aware of these tactics and will have their own appraisals undertaken by conservative corporate financial teams. However, the tension of needing to find property business in order to fund supported by a very positive developer with a favourably produced appraisal may induce greater risk-taking, particularly if the developer has a track record. The initial viability of

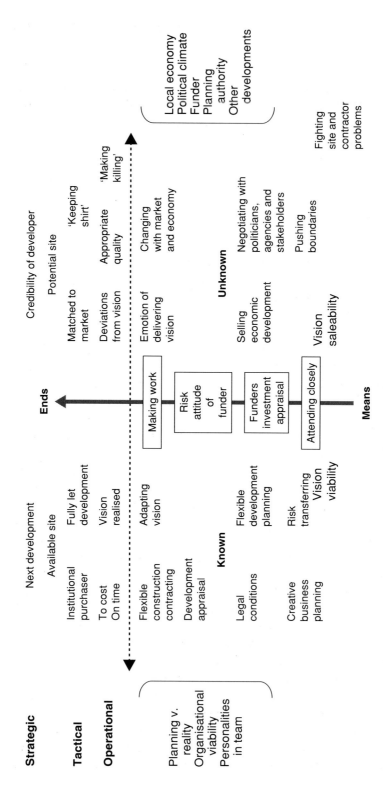

Figure 5.4 Means and ends of trader developers.

developments may be based on a vision that opportunities will arise that can be exploited to make it a success. This puts pressure on the construction, first to keep down costs in order to work with the initial appraisal and second to change the project as opportunities arise.

Modern development financing involves a creative packaging of different forms of finance. The conditions over time may be allowed to change, and financial packages may include deferral periods, draw-down facilities, rate swaps and selling on of investments. Financial packages are created substantively around the needs of the lender, balanced with a negotiation by the developer. The consequences of this financial package for others involved in development are rarely considered. Experienced developers will manage the funders and the interaction between the financial arrangements and the development arrangements. It is important for the trader to maintain knowledge of construction progress in order to protect the financial arrangements.

The trader makes more profit by getting the site developed and occupied early. The balance is between undercutting the market to achieve occupancy so as to cut potential losses and to encourage others to lease property, and waiting for a better tenant. Thus, the trader developer's view of time changes as the scheme reaches its maturity point, when it will be returned to the funder if it is not fully let. The trader may have to pick up the leases on the un-let property. The trader may pass on some of this risk to construction without them being in control over it by, for example, delaying payment or find-ing reasons to refuse completion. This is not trivial and may not be apparent, as some of the financial arrangements will be commercially confidential. This may induce suspicion and lack of trust, as decisions are taken that make the development arrangements more difficult and onerous.

Most funders keep a close watch on the development and con-strain the developer's decisions. Their objectives are different, and this conflict on perspective of what is required may surface. Both the funder and the developer share the objective of a fully let develop-ment, but the conditions of this are different. When things go wrong, it is unlikely that the institutional lender will immediately call in their security, as the transaction costs of this are too high. They would prefer to see the development completed and successful so that they can recover more money. They will be monitoring a development and will restructure the finance and even assist with management to ensure that success happens on their terms. This will involve greater charges on the project and the lender taking a greater equity stake so that they recover the value of their intervention. For developers, this is a backstop, but it provides a fuzzy area around which they can 'keep their shirt'.

The availability of good sites is limited, and thus opportunities need to be spotted or ways of overcoming problems need to be devised that others have not seen. All sites then face competition, and paying too much for a site can limit the quality of development that is worthwhile and make failure more likely. The initiation of development involves finding a site, seeing its potential, getting ownership (or assembling ownership), getting its potential visualised, creating a business plan and getting finance. Sometimes this can take years and persistence under changing circumstances, which can pay off when a number of factors come together to create viability. The secrecy about this and the implications of site problems flow through into the design and construction but are often hidden.

Trader developers make clever use of their knowledge of the market, trying to ensure that the development is in the right place and at the right time. Their attention is focused completely on the development, which enables them to foresee dangers and opportunities. Being small, they can change direction quickly without a long procedural and calculative justification. Most trader developers have good business connections in the area in which they are developing. Their decision-making is a mixture of intuition and calculation: intuitive for making the scheme work and making their profit, but calculative to demonstrate to the funder the initial viability and the ongoing viability of their vision. This sets up a paradox that infuses their approaches and may be the instigator of changes in action. They need to be aware of the position of the funder being able to justify action to them in their terms, to understand the limits of deviation, and to protect the funder from problems. One contentious area is the approach to regulation: an institutional funder has to be seen to be fully compliant, whereas the trader can negotiate and pressurise authorities to get what it wants and within its timescale. Trader developers need to be good at selling their vision against opposition, even when things are not looking good, and to negotiate effectively with the vision. They will employ consultants to strengthen their arguments and mitigate risk. They need to use their time well, as this constraint affects their profit; therefore, they need to make things happen in time by motivating and cudgelling people. They are able and willing to accept the risk of making decisions when others (including the funder) are unwilling or unable to make a decision.

With regard to the building, the trader developer sees it principally as a functional space that needs to be lettable. The specification of this value is a complex process full of conflict. There may be the institutional lender's specifications to be met. There may be tax advantages to be exploited, which may need to be hidden through an equivocation of purpose of the building. Other aspects of the building, however, may make its value higher, such as aesthetic or historical renown, finish quality, good spatial form, and adjacency to other amenities, such as transport or natural features. These features may allow the building to be let more easily, but this also increases cost. If this added value can be offset with higher rents, then features are good, but rental values for building types are set in a location, which means that only marginally higher rents can really be achieved.

The trader developer needs to be understood as an individual or group of people working for individuals. The differences in the form of finance have a great effect on the way in which decisions are made and the way in which stakeholders behave in the face of events. This puts pressure on the construction economics and time framework. During a development, however, the trader developer's attention may be distracted by the next development. The trader developer's longer-term thinking is directed towards the next development, and the developer is always looking for opportunities. This can cause difficulties, as it interferes with their decisions. The trader developer manages many contradictions between the funder and the development. The industry can assist with this, which helps them to be successful and secures a better operating relationship beyond contract. They are not apprehensive about contradictions or the unknown but use them as opportunities to enhance the development.

Mixed-use regeneration developer

Mixed-use regeneration developers are opportunistic like straight trader developers but need to be more creative and flexible. In many ways these projects are larger, much riskier, and have requirements that change during the development. For mixed-use regeneration developers, the building is more critical, as it needs a strong identity around which the area regeneration takes place and that can accommodate changes of mix. Figure 5.5 shows the means and ends of regeneration developers.

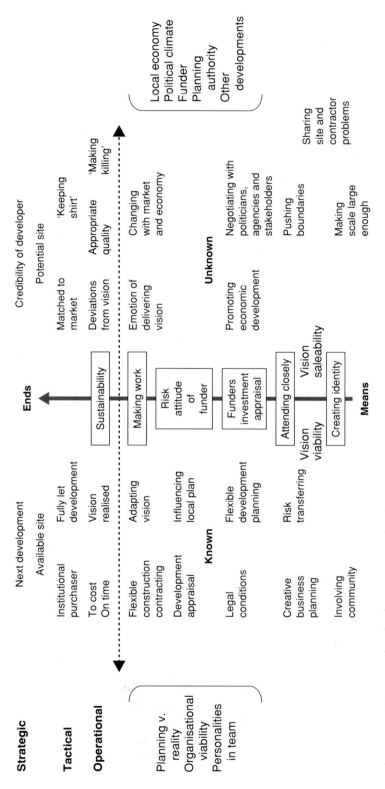

Figure 5.5 Means and ends of regeneration developers.

In the case of mixed-use regeneration, the development appraisal is even more tenuous. In this case, the local conditions current at the time of the appraisal, when the site is un-regenerated, would indicate extremely low returns. Thus, the historical data are not valid. The objective of regeneration is to make an area rise in desirability and thus command higher rents. The added risk is that this will not work. Thus, the sustainability of the scheme is of critical importance, and the developer will provide a social development aspect, such as social housing or community facilities, that will make the development more attractive. Indeed, the financial viability is not dependent only on the development itself but on the elevation of the whole area surrounding it. This requires intervention by others, including local authorities but possibly also other developers. One issue is, then, whether the development is large enough to facilitate the area regeneration. This regeneration can take time and is dependent on the propagation of the vision of the development. Indeed, a difficulty is that to do this requires a comprehensive redevelopment. It is less possible to undertake a phased scheme as the small start will not induce the desired positive feel for an area. All this requires an adequate package of land in the right place to get regeneration started. Land assembly can be difficult and time-consuming, although regeneration companies and English Partnership may be willing to use government money and compulsory purchase to make this happen. The disadvantage of this, however, is that public-sector-induced projects require particular content, and also the conditions on accountability may make the development less attractive.

Thus, the degree of the unknown is the key characteristic of regeneration schemes, and this induces particular approaches to design and construction. The developers have a clear outlook of how they wish their risk to be passed on. They need both consultants and contractors to take on this risk, and this can create tension within the project. They expect everyone to be as personally committed as they are. These developers monitor everything with close attention and personal involvement. They anticipate hidden problems. When things go wrong, everyone is expected to join in to find a solution; the developers will fight their way out with full effort. If organisations are hiding problems, then the whole development is put at greater risk. Organisations are expected to help deliver success rather than be contractual, and this success can allow greater payments. This has to be done at risk and through trust by the consultants and contractors.

The unknown is what regeneration developers make money from, by being able to be flexible and move quickly when opportunities arise. Time is longer for regeneration developers than for trader developers, as regeneration needs to wait for the increased value as their development affects an area. Therefore, their thinking can be more open and they seek to be creative. They see opportunity where others see risk. Decision-making is intense and emotional but understands the professional approach; however, regeneration developers are willing to think outside the box. Thus, there is a paradox in regeneration development between a flexibility of opportunity and a security of delivery. Regeneration developers maintain rationality in their concept of the whole of the project, which is aligned to the funder, but non-rationality in the delivery of parts of deals.

5.6   KEY ISSUES

❏ Property companies, trader developers and some small developers are highly experienced clients with high expectations and beliefs about how construction should take place.
❏ Good land opportunities are scarce and will be fought over, and thus there is a degree of secrecy about developments.
❏ Property companies are bureaucratic, having a functional structure and rational procedures involving deliberate quantifiable decisions undertaken in committees.
❏ Property companies are more interested in the way in which a development fits their portfolio rather than the development itself.
❏ Property companies see buildings as commodities that are distant from the development problems, particularly the local aspects of the project.
❏ Property companies may lack understanding of their own decision-making across their differentiated structure.
❏ Property companies have financial strength but want certainty.
❏ Currently, funders are in competition to lend to experienced traders, even with marginal schemes. This causes a contradiction between funder and developer.
❏ Traders need to maintain the relationship with the funder, thus they need to oscillate between being speculative and being corporate.

❏ Traders need consultants and contractors to work closely with them in order to give developments credibility, protect them from problems and allow flexibility.

❏ Regeneration involves an evolution in detail around a strong vision. Thus, change is to be expected.

❏ Regeneration requires scale, some iconic design, and heavy marketing of the vision in order to convince leasers that the development is worth buying into.

REFERENCES

BPF (2004) *Property in Perspective 2004: Why Should We be Interested in Commercial Property*. London: British Property Federation.

Cahill, K. (2002) *Who Owns Britain?* London: Cannongate.

Cullingworth, B. and Nadin, V. (2002) *Town and Country Planning in the UK*, 13th edn. London: Routledge.

IPF (2003) *The Size of the UK Commercial Property Market*. London: Investment Property Forum.

IPF (2005) *Understanding Commercial Property Investment: A Guide for Financial Advisers – 2005 Edition*. London: Investment Property Forum.

Isaac, D. (1994) *Property Finance*. Basingstoke: Macmillan.

Isaac, D. (1996) *Property Development: Appraisal and Finance*. Basingstoke: Macmillan.

Kloman, H.F. (2000) An iconoclastic view of risk. Presented at the AIRMIC Convention 2000, Gleneagles, Scotland, 2–3 November 2000.

Land Securities (2004) Annual Report. London: Land Securities.

MacGregor, B. and Hoesli, M. (2000) *Property Investment: Principles and Practice of Portfolio Management*. London: Longman.

Newman, M. (1997) *Marketing in Commercial Property*. London: Estates Gazette.

Rosenheim, J.M. (1998) *The Emergence of a Ruling Order: English Landed Society 1650–1750*. London: Longman.

RESOURCES

Ball, M., Lizieri, C. and MacGregor, B. (1998) *The Economics of Commercial Property Markets*. London: Routledge.

Millington, A. (2000) *Property Development*. London: Estates Gazette.

Ratcliffe, J. and Stubbs, M. (2003) *Urban Planning and Real Estate Development*. London: Spon.

Warren, M. (2000) *Economic Analysis for Property and Business*. Oxford: Butterworth-Heinemann.

Accessible Retail: www.accessibleretail.co.uk
British Council for Shopping Centres: www.bcsc.org.uk
British Council of Offices: www.bco.org.uk
British Property Federation: www.bpf.org.uk
Cambridge International Land Institute: www.cili.org.uk
Investment Property Databank: www.ipdindex.co.uk
Investment Property Forum: www.ipf.org.uk
Society of Property Researchers: www.sprweb.co.uk
Urban Land Institute: www.uli.org

6 Supermarkets as Clients

6.1 INTRODUCTION

Supermarkets and grocery retailing is one of the most successful sectors of the economy in the UK. It was estimated in 2003 to be worth £103.5bn (Euromonitor, 2004), but it is its continuous rise of about 5% annually that drives a positive aura through the market. It is a sector that is dominated by large retailers, with four – Tesco, ASDA–Wal-Mart, Sainsbury's and Morrisons–Safeway – commanding 83% of value sales in 2003. Tesco is the market leader, with about 30% of UK food retailing, a turnover of £21.6bn and a pre-tax profit of £1.3bn (Reuters, 2005). There are a number of smaller players that command specialist markets (low price, quality, class, convenience) and a number of regional and local players with historical ties in particular areas. It is the rise and expansion of the big four, however, that characterises the sector.

The stores of the big four are located throughout the country, although the dominant retailer differs in each region, with ASDA and Morrisons in the north and Sainsbury's in the South (CACI, 2004). This feature of this business market dictates that there are a number of battlegrounds where dominance in the market will be fought for strategically. ASDA–Wal-Mart is part of the world's largest retailer, in a much smaller way, however, Tesco is working internationally with stores in a number of places in Europe and Asia and, in 2005, taking a share in a US company.

Supermarkets have been in existence for over 150 years, beginning with the emergence of multiple stores in the 1850s. These multiple stores formed over half of grocery outlets in the early years of the twentieth century and created new methods of retailing, with brightly lit displays, price ticketing, advertising and national branding. Alongside this was a consumer- and community-led cooperative movement that placed (and still does place) different commercial value in its enterprises and made multiple stores universally acceptable. The emergence of bigger chain stores in the USA was replicated in the UK with the merger of several grocery firms to form International Tea. Following the Second World War, a shortage of cheap labour encouraged the development of self-service. This, and the developing theories of retailing, determined that economies of scale brought greater profits and allowed a market-oriented approach to reducing prices. Thus, supermarkets spread from a few hundred in the 1950s, to around 2000 in the mid 1960s, to being the dominant outlet (80%) of grocery sales in the 1990s (Seth & Randall, 2001).

6.2   THE BUSINESS ENVIRONMENT: STRATEGY IN THE WORLD

Supermarkets are classic traders, buying goods and selling them on at a profit. In this they do no production and little transformation but add value through delivery to their location and economy of purchase. However, the powerful position of the big four provides them with the ability to control their market through marketing, branding, research, publicity campaigns, advertising and promotion, and to control their supply chain through strong procurement practices, proactive management and training of suppliers. They can be modelled effectively through business-system hierarchy theory, which comprises strategic, tactical and operational domains (see Section 3.3). A map of the business environment of a supermarket business is shown in Figure 6.1. This map shows areas relating to market dominance, supply-chain management and business operations, which relate to strategic, tactical and operational concerns.

Strategically, supermarkets have become powerful national and local political players, the former as business leaders and the latter through creating economic development and employment. Thus, not only does the external world affect the big supermarkets, but also the big supermarkets have much power over the external world. The range of goods in which the supermarkets trade is increasing: supermarkets are already dominating the sale of petrol and are now

LA, local authority.

Figure 6.1   Issues and actors in supermarket business.

venturing into services such as finance and insurance. Their skill is not only in retailing but also in logistics, i.e. the efficient supply of goods. In order for supermarkets to operate, they have to devote considerable time and effort on their relationship with the external world. Thus, their strategic thinking is influenced heavily by these pressures, both in complying with the requirements of the external world and in seeking ways to change the world or to operate around onerous regulative requirements.

The dominance of the big players has created an oligopoly (the competition between few), which has identified the sector for attention by government because of its potential distortion of both the retail market and the suppliers market. This fear of market distortion and strong lobbying from interested parties, particularly farmers, has forced the government to conduct a series of enquiries through the Office of Fair Trading. Most notably, the Competition Commission undertook a major study, which it reported in 2000 (Competition Commission, 2000). This report revealed a considerable amount of information about the strategy and operation of the supermarkets. Although generally viewing the companies as operating satisfactorily, it did identify problems on the retail side of below-cost selling, price-flexing (raising prices in the light of local competitive conditions) and focus on the price of a small number of products. However, it was on the buying and supply-chain side that much of the lobbying for the study had been emanating. The study considered, therefore, that the power of the supermarkets forced suppliers to contribute non-cost-related payments, to give discounts, sometimes retrospectively, to accept changes to contractual conditions without adequate notice, and to accept the transfer of risk to the supplier. These supplier issues were dealt with by the adoption of a code of practice, which the supermarkets had a statutory duty to sign and comply with. The most significant development in the dominant structure of the sector in the past few years was the purchase of Safeway by Morrisons. This was referred to the Monopolies Commission, and part of its judgement was to force the sale of some Safeway stores to competitors. Second to this has been the rise in convenience stores controlled by the big four. The strategy appears to be to secure total coverage in areas so that the brand is dominant.

The differentiation of the supermarkets is a strategic issue for them all. It is achieved by some of the smaller companies, e.g. Waitrose for quality and environment but not price, and Lidl, Aldi and Netto for price. One of the successes of Tesco has been its ability to span all retail markets, from low cost to high cost: all social classes shop at Tesco, from A1 to E. Other supermarkets have not achieved this extent of customer base and are in a dilemma about whether to aim for it. The creation of the supermarket as a national brand to which customers have loyalty is a major change in shopping in the past 20 years. Supermarkets now aim for brand dominance, and this has certainly been achieved by Tesco, which has been enhanced by the variety of formats of store that they now possess, from hypermarkets to convenience stores. Alongside this, however, each is trying to convey a character that is beyond their aggressive business image, involving them in supporting civil society and the community. Thus, supermarkets are actively involved in national campaigns and charities, environmental issues, infrastructure development, community support and employment provision, as justification of their political-economic position, as public relations for sales, to avoid social opposition to the business activities, and to provide a reason for customer loyalty. This works from board level to shop level and thus is a unifying activity of the business and helps to establish an identity externally and internally to the business. However, this is a complex position, made easier only by the fact the companies are all so big and with large public relations budgets. They are selling themselves to themselves as well as to the world. Most run national campaigns targeted at providing education, e.g. supplying computers or sports equipment to schools. Finally, all are concerned explicitly for the environment and have sustainability policies and clear examples of good practice. Many have amalgamated these activities around a corporate social responsibility policy; the operation of this is variable across the sector, however, and supermarkets are often accused of merely doing it for favourable publicity.

The sectors from which the supermarkets buy are mainly producers – primary producers, such as farmers, and secondary producers, such as biscuit manufacturers. The strength of manufacturers has been substantially diminished by retailers, and even well-known brands find it difficult to defend themselves against the larger supermarkets. This tension has meant that larger manufacturers are seeking ways to re-establish their positions, although this has not yet been achieved. With regard to smaller manufacturers, the multiples are presenting themselves as modernisers of practice, both in improving quality and increasing efficiency, thus making these more sustainable (Opie, 2005). The Competition Commission report suggests that there is much use of supermarket power to control suppliers for the benefit of the supermarket.

The large multiples do not need trade associations. They have enough power and influence to talk to government and agencies without having to present a wider front. They can afford good lawyers and consultants to write reports that present their cases. Smaller supermarkets, however, find an association important to engage with lobbying and to present a unified position against impending legislation.

All the major supermarkets fund development from revenue. This allows them to be not beholden to financiers who induce conditions that may not be effective for the business merely for the investment. All are publicly quoted companies, however, and as such report results to the stock market. This reporting cycle becomes a major time horizon at corporate and strategic level, which has consequences for the unfolding tactical development and the reporting of store performance. In addition, there is a growing requirement for corporate governance, which demands a more transparent approach to business. Such activities are

This has had a consequence on the development and structure of stores around the category of shopping trips made by consumers, namely one-stop shopping, secondary or top-up shopping and convenience shopping. Tesco made a statement in the Competition Commission report that presented its distinction between these customer habits:

(a) *Pure one-stop shopping*. This is where all grocery items for a given period are purchased from a single store in a single visit, which may or may not be a supermarket.
(b) *Mixed one-stop and top-up shopping*. This is where most grocery items for a given period are purchased from a single store in a single visit, with top-up items purchased during another visit to a supermarket (which may be the same one as before or a different one) or to another (non-supermarket) store.
(c) *Sub-basket switching*. This is where most grocery items for a given period are purchased from a single store in a single visit. However, a specific set of items will be bought from a different store, which may or may not be a supermarket. These items are not top-ups but are products that the customer would prefer to purchase from another retailer.
(d) *Multi-shopping*. This is where the required grocery items for a given period are purchased during a number of different visits, which may be to the same store or to different stores.

Competition Commission (2000)

Shopping behaviour is changing and is influenced by both social and work pressures of customers and planning policy. Thus, one-stop shops are stores in excess of 15,000 sq. ft; these have been getting bigger, however, and stores in excess of 50,000 sq. ft are common, with some even at 100,000 sq. ft. The economics of these hypermarkets are significantly different from smaller stores, as is their store management, which becomes one of operating a substantive business with sometimes £40m p.a. turnover and employing 20 to 30 managers and 350 staff. At the other end, the big four have entered the convenience store market, which caters for convenience, emergency and impulse food shopping. These stores are less than 3000 sq. ft and are located in town centres, railway stations and petrol stations. The Office of Fair Trading considered that these stores were in a different market and so did not constitute an unfair monopoly (OFT, 2005).

With regard to the business competition, apart from customers, who to some extent are loyal, then the site of the competition between supermarkets is the supply chain (Harvey, 1999). Tesco's success is the fact that it can get its supply chain to deliver and operate cheaper than other supermarkets. This supply-chain competition includes not only the cost of the goods but also the logistics of delivery, which has reached high levels of efficiency, but the supermarkets are researching new developments in this area to reduce costs further. The cost of supplying groceries comprises: cost of goods (83%), staff costs (9%), capital costs (3%) and other operating costs (5%), broken down into store costs (75%), distribution costs (13%) and overheads (12%) (Competition Commission, 2000). Thus, the tactical importance of supply-chain management can reduce the cost of goods, the cost of distribution and the effectiveness of distribution and allows the supermarkets to have the right quantity of the right products on the shelf at the right time.

6.4   BUSINESS OPERATION

Supermarkets are fiercely competitive businesses. The market dominance by Tesco means that the other supermarkets are trying to achieve the performance of Tesco, often by blatantly adopting Tesco's processes. Thus, strategically, businesses are trying to differentiate themselves from Tesco in order to compete in the marketplace while at the same time copying Tesco in its processes. Tesco, on the other hand, is searching for ways to keep ahead of the field. This competitive drive induces much of the business practices and also the way in which supermarket organisations engage with and experience the construction industry.

The competition is to achieve a higher percentage of the national retail business market. This means working to improve current business practices while developing new opportunities either in locations where the organisation has less market share or in new business ventures such as non-food and convenience stores. One of the big competitive issues is price of goods at point of sale, although this may focus only on certain staples. All of the major retailers promote low cost and high value. ASDA has this as its major mission statement; Sainsbury's has now adopted it, but previously it had more of a quality promotion. The supermarkets' management are always demanding lower cost at the point of sale, and this is an important driver of business. However, this means that this focus on volume of business is outweighed to some extent by the basic need for margin from that business, and so there is a degree of tactical conflict and a degree of playing with prices.

The prime target for sales is the customer. In the past, price and attractiveness were central to sales. Now, however, a deep understanding of customers' needs and habits has been achieved. Supermarkets are in a business whose prime purpose is to get customers to buy its goods; this is not sufficient, however, and the key is to deliver a customer experience. The belief is that it is the customer's experience that induces the customer to buy goods and to return to buy more. The tactical dilemma, then, is how much to spend on enhancing the experience and how much to spend on inducing sales through price or promotion. The customer experience might include store environment and accessibility, but it also includes the provision of choice and availability of goods. Thus, most supermarkets are developing a wider range of goods for sale and driving changing consumption, such as convenience dishes, fresh bread, cooked chicken and organic ranges, as well as offering additional services, such as pharmacies, opticians, dry-cleaning, photo-processing and cafeterias.

As part of the supermarkets' desire to offer goods at a lower price, they have had technical innovations, mainly in information technology (IT), such as electronic point of sale (EPOS) and electronic data interchange (EDI) with suppliers, self-scanning, transport routing, energy-saving (particularly refrigeration) and home shopping. These technical advances have affected mainly the background operation of stores but also the generation of management information, which is now at the heart of supermarket operations. These information systems allow senior management to compare information from the calculative business planning of stores with the weekly operational performance and to use this to control the hierarchical corporate structure. Stores have a number of key performance indicators, including sales per square metre, costs (particularly labour), store profitability, stock levels, human resources and customer services, which are reported on a weekly basis, with comparisons being made against the same week in the previous year and against similar stores.

As part of the management information, a cost to get an item on to the shelf is produced; this is then targeted by management to be reduced. In this, the concepts of supply-chain management and logistics are utilised, both to improve products and to provide them at lower costs or higher margins at the shelf. This calculative world and belief in control is characteristic of both supply-chain management and logistics and has delivered considerable reduction in cost. There are many criticisms of this in the media, as often it is seen merely as getting suppliers to accept lower and lower payments for the goods and to be willing to accept the retailer-induced changes at the supplier's cost. However, logistics management has had considerable success by being both lean, (i.e. removal of waste from the system) and agile (i.e. reduction of lead times in the delivery of goods). The advent of real-time sales information from stores now allows the supply process to be pulled from sales, thus making the process more efficient and more timely. This requires good product coding and identification throughout the product's journey from producer to shelf and also an integrated management system that intelligently initiates the supply and controls the flow to the point of sale.

The acquisition of customer knowledge has become a science and commands a great deal of the supermarkets' strategic, tactical and operational attention. The information base of regional and local demographics has been enhanced by specific customer details of purchases at point of sales, which is tied to each customer individually through loyalty-card schemes. These data define the quantity and characteristics of the customer base around a store or potential store location and become the cornerstone of the rational planning of stores. This can determine the business plan for the potential income and thus what it is worth spending on a particular store. In addition, it determines the mix of the offer at the store, i.e. the range of goods that will sell most and the space that should be allocated to these goods. At the next level, such modelling allows the most efficient location of distribution centres being able to predict the frequency of delivery vehicles and the necessary form of layout to cope with this load.

The operational management at a store is dominated by the general manager. Large stores have a general store manager, one or two deputies and a number of other senior personnel, who direct and operate the store almost as an individual business entity. Beneath these are the retail commodity managers – e.g. dairy, vegetables, clothing, bakery, delicatessen – depending on the offer of the particular store. Beneath this are floor managers, who act as a presence to the customers and the staff. At the other end, there is the management of store personnel who tend to work part-time, short-term and in relatively unskilled jobs. This again requires a clear hierarchical structure and clearly set-down procedures. In this environment, targets are set at all levels, i.e. corporate, division, stores, departments and individual personnel, and these are monitored and reported on. This highly rationalised approach utilises numbers of people and time of all staff. The management system believes in this formalised approach and the rewards it delivers in the sense of monitoring and control. This arrangement induces a structured corporate approach to management and the career path that must be followed.

There is a conflict in supermarkets about this scientific knowledge, with rational planning versus store management, who believe that they really understand intuitively a local customer. Although strategic and tactical decisions are made in this highly calculative way, most supermarkets give some discretion to local general managers, both as a motivational aid and as a more precise alignment to the market.

6.5   EXPERIENCE OF BUILDING: FROM UNKNOWNS AND CONTRADICTIONS TO MEANS AND ENDS

The business power of supermarkets makes them have a firm belief in their ability to control the world through their plans; in this sense, they are rational, goal-centred bureaucracies with employees working through roles in order to meet the corporate objectives. This attitude pervades a supermarket's approach to the building and to building, where it is also seeking to be in control and will pay for this privilege. The way in which a supermarket uses its knowledge and process, and the way in which it handles gaps and contradictions that come out of this, are presented in Figure 6.3.

Supermarkets have a strategic problem of needing to be more successful. They need to keep expanding, being more profitable and being more efficient. Their employees are in this treadmill of success but need to search for more success. Their difficulty is to know where to go next and, for the individual, to know how to make a name for themselves. As the businesses become bigger, they become more bureaucratic. The corporate structure of the organisations means that there is a contradiction between the corporate view and the local store view. The boundary between the store as the primary delivery mechanism and the central organisation becomes bigger. Thus, investment might be directed overseas, leaving underinvestment at store level but an expectation of performance, even without investment. In addition, the standardisation resulting from the rationalised corporate decision-making can produce anomalies at local level. The first of these will be hidden and apparent only from the behaviour and expectations

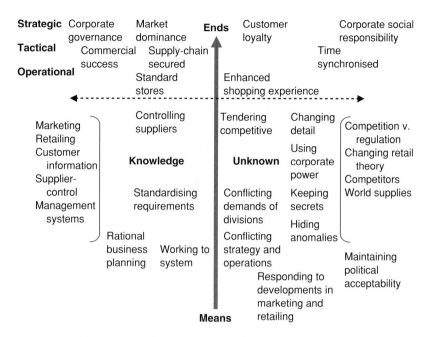

Figure 6.3 Means and ends of supermarkets.

of corporate management. The latter is hidden at corporate level but is known at store level, although it may have to be hidden from or not conveyed to junior store staff.

Theories behind retailing and merchandising are developing rapidly. As one theory becomes the latest fashion in a supermarket chain, it is difficult for other chains not to adopt it as well. Thus, there is an environment/climate of change constantly around tactical approaches to sales and sales spaces. The drive for operational efficiency and reduced cost has long-term consequences of supplier viability, product quality, product variety and human cost. This contradiction is latent, but also the consequences of the drive are hidden as they emerge. There is a fundamental problem with such structured operational systems: individuals feel constrained by them, and the systems do not respond to customer needs on the ground. Thus, the system becomes the reason for being rather than the objective of the system. The more that is invested in systems, the more strongly management seeks the organisation to adhere to them. All systems have problems and errors, and thus the perfunctory response from unmotivated staff is to do what the system says, even if it is wrong. This induces people to be defensive and, alongside strong hierarchical managerial practices, may reward adherence to the system rather than enable the system to develop.

The building itself is a dilemma and involves a continuing debate, disagreement and compromises, which can be at odds with the organisation. The prime purpose is retailing, and that is what is valued and rewarded. The building is almost an annoyance, as it constrains the retailing and takes away valuable resources from this primary activity. Against this is the awareness that the supermarket is selling a retail experience that is not only goods and services but also an external and internal environment. Thus, the external environment is important, not only for easy access but also as a pleasant addition to a community in an urban setting or to a landscape in a rural setting; for example, ease and quantity of carparking are in conflict with the visual quality of the setting. This is extended to the building itself, where visual quality may get a building known and differentiate it from a competitor, but it does not sell goods. Inside the building, there is a dilemma between creating the operational form for efficiency or for the quality of the customer experience. There is also the contradiction between non-retail areas such as delivery and storage space and retail areas. Retail sales require the goods to be put on the shelves, but this activity is a lower value to sales space itself. Other issues about space and customers concern checkouts. Waiting is clearly an aspect of customer experience, but again the provision of more checkouts reduces store space for goods, although long queues also prevent shoppers accessing the shelves.

The financing of development generally comes from the revenue of successful stores. Clearly, stores that are less successful have a greater job in justifying remodelling and development expenditure to their shareholders. This dilemma of shareholders is endemic in all publicly quoted companies and induces a further short-term view of delivery to annual cycles of results and dividends. In a competitive world, the provision of extras such as external infrastructure improvements becomes a great drain on profits. These are strategic negotiating weapons that allow access to difficult locations and the ability to develop with less constraint or battles. This is also part of a wider public relations activity that involves national campaigns to local community prizes. Public relations is required to avoid a concerted attack by the political establishment.

These gaps and contradictions in a supermarket's knowledge and processes determine its approach to building. Business unknowns come out of the environment but also from the contradictions inherent in practice within an organisation. During times of change, these contradictions are exposed; the change induces opportunities for individuals and sections to manipulate the situation to their own advantage. The attitude to time causes a variation in response to the project. Time horizons include a five-year store development on a regional basis, with the expectation that the construction cycle lasts about a year for a new store. The remodelling cycle takes place every four to five years but can be induced earlier by local competition. The refit cycle occurs every two to three years. Both remodelling and refitting require working in an active store, which forces construction to be undertaken against a customer cycle. The store and head office get nervous as the company nears its financial reporting points, which is annual but with six-month returns. Decisions at these reporting times may be problematic. They believe that the construction industry is just another supplier and that what applies to a grocery supplier can be made to apply to building supply; thus, the construction industry is expected to be flexible and compliant.

6.6 KEY ISSUES

The supermarket sector can be summarised as follows:

❑ Supermarkets are powerful clients that believe in their position in the world. They seek market dominance.
❑ Supermarkets are corporate bureaucracies with business conflicts between head office and stores.
❑ Supermarkets sell a retail experience as much as goods.
❑ Supermarkets compete on supply-chain costs and delivery.
❑ Supermarkets have high expectations of service from suppliers, and expect flexibility at no cost to them.
❑ The industry needs to negotiate the balance between the delivery of expectation and reward.
❑ Store development exposes problems in the relationship with customers, communities, local small businesses and local regulators, which can affect the projects.
❑ Undertaking projects in working stores, there are problems with customers, staff and the protection of goods.
❑ Success is a compromise between satisfying the corporate organisation and satisfying the individuals in the local team.

REFERENCES

CACI (2004) Life after Safeway. Press release 6 December 2004. London: CACI.

Competition Commission (2000) Supermarkets: A Report on the Supply of Groceries from Multiple Stores in the United Kingdom. London: Competition Commission.

Euromonitor (2004) Grocery Stores, Food Retailers and Supermarkets in the UK: Executive Summary Report. London: Euromonitor International.

Harvey, M. (1999) Innovation and Competition in UK Supermarkets. CRIC briefing paper 3. Manchester: Centre for Research on Innovation and Competition, University of Manchester. http://les1.man.ac.uk/cric/Pdfs/bp3.pdf.

OFT (2005) *Supermarkets: The Code of Practice and other Competition Issues – Conclusions.* London: Office of Fair Trading.

Opie, A. (2005) Supermarkets are driving standards higher. *The Edge*, **19**, 28–29.

Reuters (2005) Tesco's market share slips. *Reuters Business News*, 17 November 2005.

Seth, A. and Randall, G. (2001) *The Grocers: The Rise and Rise of the Supermarket Chains.* London: Kogan Page.

RESOURCES

Fernie, J., Moore, C. and Fernie, S. (2003) *Principles of Retailing.* Amsterdam: Butterworth-Heinemann.

Reynolds, J., Bell, R., Cuthbertson, C., Cuthbertson, R., Davies, R., Dragan, D. and Howard, E. (2003) *Retail Strategy: The View from the Bridge.* Oxford: Elsevier Butterworth-Heinemann.

Association of Convenience Stores: www.thelocalshop.com. Formed in 1995 as the retail trade association for the UK convenience store sector and representing over 32,000 shops.

British Retail Consortium: www.brc.org.uk. Lead trade association representing the whole range of retailers, from large multiples and department stores, through to independents selling a wide selection of products through centre-of-town, out-of-town, rural and virtual stores.

Institute of Grocery Distribution: www.igd.org.uk. Key research organisation for the sector; provides a forum for discussion, learning and opportunities for improvement and the development and sharing of best practice.

Tesco's 'My Property': http://myproperty.uk.ocset.com. Tesco's Property Knowledge Management System available to its supply chain for land acquisition, to maintenance developed by Styles and Wood.

Styles and Wood: www.stylesandwood.co.uk. Leading provider of retail property services.

7   NHS Acute Trusts as Clients

7.1   INTRODUCTION

The National Health Service (NHS) was set up in 1948 and is now the largest organisation in Europe. This historical legacy has set its philosophical position on the delivery of health and has provided organisational idiosyncrasies that characterise the service today. The Second World War paved the way for a shift in social and economic position of the UK. Thus, the political position that everyone would have access to the healthcare they needed, regardless of income, was embodied as a right and produced a high expectation of service. Organisational thinking at the time, and as part of the experience of war, determined that a centralised planned and controlled system was created, although opposition from the medical profession (particularly regarding doctors' contracts) did not allow this to be implemented completely.

In economic terms, the NHS was created as a command solution whereby the state decided how much healthcare it could afford and what this was going to be spent on. The NHS owns most of its infrastructure and employs most of its people. The state does this through taxation, thus giving people no choice on whether to pay or how much to pay for health. With this prescribed offering, however, healthcare is free, apart from charges on items such as dentistry and prescriptions.

Health is a complex concept in its own right, and this problem pervades the service. Everyone dies, and we relate health to an expectation of a norm. This norm is often defined by statistics, so that mortality rates, a well-known and calculable metric, are used to compare countries and developments over time. A person born in 1841 in England or Wales could expect to live to only just over 40 years, but by 1998 a man's life expectancy had increased to 75.1 years and a woman's to 80 years (OHE, 2000). These changes were due mainly to reductions in infectious diseases carried by air, water and food as a result not only of health-service improvements but also improvements in infrastructure (e.g. water supply), work environment and food. A health service is required to do more than just offset death. It must also improve an individual's quality of life, i.e. their being able to carry on life as normal without impairment, pain or discomfort. Indeed, much of modern medicine is about trying to achieve this. The problem, however, is that the norm of health is not static, and as health improves so our expectations of health are higher. At the same time, our ability to make people healthier is also expanding with experience and developments in drugs, medical technology and medical procedures.

For an individual, health and being healthy are not about statistics but are concerned with a feeling of wellbeing. This is bound into other aspects of society, such as self-worth, and these are not physical but psychological attributes. This links with shifts in the concepts of health from being only about the body to also being about mental illness, where newly defined conditions such as stress need to be treated. Health then is not only about fixing a broken machine but about supporting individuals within their family and social groups to live the lives that they expect to lead.

This is probably the most highly emotional, highly emotive and highly political aspect of our society, which means it commands enormous attention by all. The political and economic dimensions range from the individual, through business and the economy, to society as a whole. Everyone has an opinion, and the debate encompasses philosophy and ethics as well as economics and politics. Two issues exemplify the wider problem – efficiency and equity (OHE, 2000). These both present paradoxes that do not have a resolution, so they can only be managed.

Equity is based on an idea of social justice, which is bound up with ethics. Horizontal equity is concerned with the equal treatment of equal need. This means that to be horizontally equitable, the healthcare-allocation system must treat two individuals with the same complaint in an identical way. Vertical equity is concerned with the extent to which individuals who are unequal should be treated differently. Thus, should more treatment be delivered to those with serious conditions than to those with trivial complaints? The utilitarian position is to benefit the largest number. This presents itself not only as a decision about whether to spend more on primary care (greater number) than secondary care (special problems) but also whether we should spend more on social causes of ill health, such as poverty, housing, poor diet, lack of education and stress, rather than on medical aspects of health.

As well as questions about what the health service provides to individuals, there is also the question about the way in which it provides this. As individuals, we are in receipt of a service; and so what rights do we have in the way of individualised care plans, privacy, quality of environment and need for involvement? This interacts with the issue of efficiency, as it uses resources that could be used for treating others. The debate also involves issues about efficacy of service, in that providing these may make care more effective.

At the other side is the level of responsibility that society should have for individuals, or whether individuals should look after themselves. Is society interfering in the individual, or is the individual abusing society by taking too much from it? Thus, some would say we do not only have a right to health but have a duty to live a healthy life (O'Neill, 2002), and therefore we might not treat people who abuse themselves; however, the issue is also who should make the decision and how should this decision be made?

Efficiency is the second big issue. This is an economic concept concerning the maximum output from the use of resources. The problem is that health can devour resources, such that the difficult dilemma is what amount of society's resources should be directed to health and what should society expect for this? As technology improves, as new drugs are produced and as procedures are developed, then what can be done is expanding. Rather than simply putting more resources into this expanding sink, there is a belief that we can make better use of the resources. Currently the UK spends about 8.6% of its gross domestic product (GDP; of which 1.4% is private care) on health, with a targeted increase to 9.2% in 2007–08 (HM Treasury, 2004). This is less than many other countries in Europe and significantly less than the USA, where the figure is over 13% (Emmerson *et al.*, 2002). Of course, increasing a country's GDP allows it to spend more on health. In 2005–06 about £76.4bn (25.4% of budget) was spent on NHS health in England compared with £30.8bn (10.2%) on defence. Of this £76.4bn, £72.2bn was designated for revenue expenditure and £4.2bn for capital.

Over the years, the efficiency of the NHS has been criticised and periodically efforts are made to improve it. There have been both political and managerial criticisms. At a wider political level, there is the debate about whether private-sector organisations perform better than public-sector organisations. Currently there is a move to steer delivery to a more private-sector approach; the desire is not to leave healthcare to the private sector but to create the operational conditions of the private sector while retaining public ownership and accountability. Much of this move to simulating a private-sector environment is managerial, but other elements include trying to create a market, which has proved more difficult to implement. This market identifies service buyers and service providers, with complex arrangements for ensuring quality of service and value for money. More radical public–private partnership arrangements, including the private finance initiative (PFI) and buying services from private hospitals, are also being tried to ensure that a more market-oriented delivery system is in place. In addition, business-performance measurement approaches have been introduced, including extensive use of targets, benchmarks and process engineering in order to deliver a better service more efficiently. The overall health budget is being organised and assessed using programme budgeting. In this, the budget is allocated to specific objectives rather than to specific hospitals or drugs, so that success can be better evaluated and new budgets set so that they have most effect.

Over time, ideas on structure have oscillated from regional models, to local models, to national models. Currently there is more of a devolved model with a regulatory central function. This includes a high-level devolution of control over the health services in Scotland and Wales to their parliament and national assembly, respectively. The creation of trusts has been chosen by governments to implement the change from a centralised public-sector bureaucracy to a locally controlled public market corporation. A trust is a legal entity in which a body of people (the trustees) have legal control over certain property but are bound by fiduciary duty to exercise that legal control for the benefit of other people, according to the terms of the trust and the law. At the same time, the concepts of management have changed to reflect this. Where previously there was a command solution, where people simply got what was provided, the move is now to a demand-led solution. A demand solution encourages the idea of being a customer, which contains ideas about rights to choice and quality of service. This change has run through the public sector as it moves from a resource-dissemination system to an activity-costing and performance system.

This local self-managed organisational model with external regulation creates a number of tensions with resources. Expectation of service delivery is being increased continually; however, some of the resources for this have to come from increased efficiency. These efficiency increases may be more difficult to achieve, and so a significant number of trusts have gone into deficit and there are calls for tighter financial management of trusts (NAO/Audit Commission, 2005). Indeed, many of the procedures that have been put in place are an attempt to engender an awareness of costs of care to the organisations and clinicians. At the same time, because some form of rationing of care is inevitable and in order to achieve equity, the National Institute for Clinical Excellence (NICE) provides a form of rational assessment of treatments against their cost for health and clinical excellence. This is controversial both with doctors, as it displaces their clinical judgement, and with patients, as it removes public-sector therapy options from them.

The diagram in Figure 7.1 presents an overarching structure for the delivery of health as a whole in England in 2005. The names for these parts and the number of them changes as strategic reorganisations

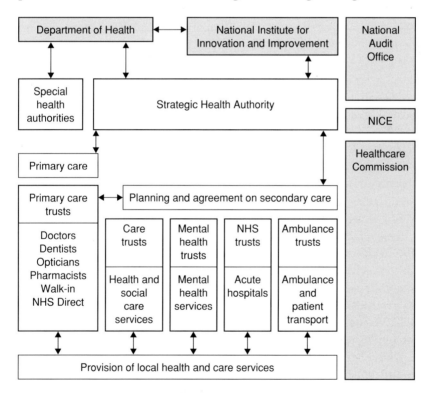

NICE, National Institute for Clinical Excellence.

Figure 7.1   National Health Service (NHS) system in 2005.

take place regularly and it is difficult to maintain a current awareness of this, even for people within the health service. What follows is only a snapshot in time and so is indicative.

The differences in the other countries of the UK are about detail and name rather than about how the NHS operates; there are different amalgamations because the service is smaller. The issue is that equity (of the same care across the country) has become an identified political objective. In 2005, England had 28 strategic health authorities (SHAs), which represent the Department of Health (DH) locally considering performance and capacity. The NHS Institute for Innovation and Improvement supports NHS clinicians and managers in their efforts to deliver improvements to their services with information and learning resources.

The big delivery distinction is between primary care and secondary care (or tertiary care). Primary care refers to the first point of entry to the service, when the patient's condition is not an emergency. Accident and emergency (A&E) treatment is located at secondary level. Primary care might be a visit to a doctor, a dentist, an optician for an eye test or a pharmacist to buy cough mixture. In 2005, these services were managed by local primary care trusts (PCTs). PCTs work with local authorities and other agencies that provide health and social care locally. As the patient-centred commissioners of care, PCTs receive 75% of the NHS budget, although most of the PCT budget goes to purchase secondary care. Secondary care involves treatments that cannot be carried out by general practitioners (GPs) and take place mainly in hospitals. These are organised and managed through NHS acute trusts, some of which are also more specialised hospitals (tertiary care), possibly even providing a national service in a clinical area.

The Healthcare Commission is an independent body that inspects the healthcare deliverers to ensure quality and value for money. It does this through collecting surveillance data from each trust, which it compares against targets set by the DH. It also compounds these to produce a star rating for each hospital. The National Audit Office (NAO), and in the past the Audit Commission, also monitors financial propriety, proper governance and stewardship in trusts.

There have been attempts to remove health from politics, but the nature of health means that it has a high constituent interest and also a high use of resources, creating conditions for continuing conflict. It can be viewed that health is so important to us individually and socially that it is influential in all aspects of society. It is so complex that the possibilities of suggesting improvements are enormous. The political pressure means that there are many suggestions about how to do it better, and one of the great characteristics of the NHS is that it is changing continually and continually being changed. This dynamic is generated both politically and managerially, with a small amount generated from developing health problems and solutions. The one thing that is constant in the NHS is change. As this complexity, conflict, instability and emotion characterise the NHS, it is not surprising that it spreads down into the NHS as a client and its approach to building. Most people in the NHS face these enormous problems everyday, and thus to understand what happens when a building takes place we need to understand how they cope with this.

7.2   THE ENVIRONMENT OF NHS ACUTE TRUSTS

The subject of this chapter is acute trusts, which offer secondary and tertiary care, although some of the thinking will apply throughout the service. Domain theory allows us to understand the way in which an NHS acute trust delivers its objectives to its environment and receives resources from its environment (see Section 3.3). Domain theory distinguishes a different purpose and a different set of achievements at political, managerial and professional operational levels. Thus, any issue or event will have an aspect that will be seen differently through each domain. The business environment of an NHS acute trust is presented in Figure 7.2. This is generalised, and local differences will exist. The internal divisions are simplified to finance, medical, nursing and information, but this will be expanded later.

Political domain

It is clear that most people conceive hospitals to be places to go if you are sick in order to be made better. This trivial operational objective belies an extremely complex institution for making this happen while at the same time fulfilling many other objectives. The difficulties of prioritisation, financial limitations, strategic position and measures of success are all initiated in a political arena. This political arena exists at both a national level and a local level, and these may not be congruent. The NHS involves people with strong beliefs and emotions at all levels. The political domain at government level is seeking to present government as aware and effective in administering the country. The full political partiality is offset by opposition and a civil service that

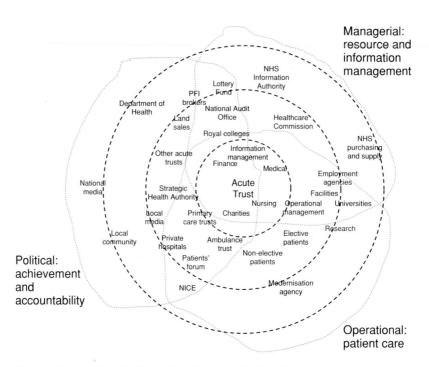

NICE, National Institute for Clinical Excellence; PFI, private finance initiative.

Figure 7.2   Business environment of a National Health Service (NHS) acute trust.

mediates action. The political opposition, however, is seeking to present its position as being superior, and thus it must try to demonstrate the inadequacy of the present government's actions. This promotion of, and attack on, policy uses advisers, researchers and academics to churn over information and find meaning that suits political purposes. As the debate occurs within the context of national events, so the meaning and direction of policy may change, which determines different priorities and creates sensitivities that react with local events. To some extent, the hospital is a vehicle of government political policy, and its successes and problems can be viewed as part of this. The 2005 priorities were cancer treatment, palliative care and treating coronary disease. In this regard, as trusts take on, for example, a PFI expansion programme or seek foundation status or use National Lottery funds, they are acting with national government policy, which can benefit the trust but also expose it to criticism.

This criticism of its actions will also have a local political dimension. Indeed, it is expected that hospitals are responsive to local needs and engage in local consultation. Thus, patients' forums and local authorities are stakeholders at the political level and can induce changes in business priorities. Some of this politics is mediated by the SHA; however, the public face of most local political action is at the hospital level. The SHA negotiations encompass the overall provision of service, e.g. whether A&E treatment is available at all hospitals. Individual acute trusts offer some of this service in competition with other trusts in the region (or even nationally), and therefore part of local politics is a lobbying for the promotion of a particular trust's position against others for funding of special services. In general, the funding stream to the acute trust from the PCTs is also subject to this competition; however, problems existing at the primary-care level can get transferred to the secondary level, e.g. a shortage of GPs may result in more patients seeking A&E care. One important aspect of this is the media. Health makes good emotive headlines, and human-interest stories of success populate the media. It is the detailed scrutiny by investigative journalists, however, that can search out problems and seek patients with complaints that can cause sudden exposures. When this happens, such events tend to be exploited by local politicians; some events escalate to the national media and national politicians. The difficulty for the trust in this situation is maintaining patient secrecy while defending its position against the media trying to catch it out.

Funding is always a political issue in the NHS, both nationally and locally. In the past, NHS trusts agreed with PCTs on the number of patients to be treated and cared for and a locally agreed cost for doing this. More than half of the money spent by PCTs went to acute services, which they previously received using a weighted-capitation resource-allocation formula that aimed to target resources at the neediest areas. In 2004 a financial-flows model was introduced, which was designed to make funding more transparent and tied directly to the achievement of outputs, called payment by results. This was intended to reward efficiency, support patient choice and encourage activity in order to reduce waiting times. Further changes involved the setting of a tariff for each procedure or inpatient spell, whether elective or non-elective. Apart from capital funding from the NHS, other funding can be achieved through private company donations, joint ventures, the Lottery Fund and charitable donations, all of which require management time to patronise, negotiate and deliver.

Since the 1990s, all large capital projects can be funded only through a PFI route, and this looks set to continue, although its mode of utilisation might change. PFI involves a specially constructed business consortium, called a special-purpose vehicle (SPV), which funds, constructs and operates the new buildings. This functions like a lease, whereby the SPV receives payments from the hospital over the life of the concession controlled by a service-level agreement. The deliberations for these PFI projects have always been tortuous, with political arguments over the justification against the public-sector comparator (Wynne, 2002), risk transfer and the legal detail of the service-level agreement (SLA) (Lefty, 2005). Such PFI projects have difficulty in maintaining the interest of the private sector because of the extended time and non-reimbursable costs of the negotiations.

One advantage of PFI to a trust is the reduction in capital charges set by the Treasury and the DH. The objectives of capital charging are to promote an awareness of the true cost of capital, improve decision-making on asset acquisition and disposal and promote efficient use of assets. The cost of capital charge (3.5% in 2005) is applied to the average net relevant assets (NRA) of the trust to determine the cost of capital for an accounting period. The cost of capital charge is applied to most assets and liabilities in the balance sheet, with liabilities attracting a negative charge (i.e. a credit). The cost of capital charge ensures an appropriate return on taxpayers' equity. PFI schemes that are funded and owned by others do not succumb to capital charges. Trusts have other mechanisms of reducing the effect of capital charges on new assets, for example by getting the district valuer to appropriately revalue the whole estate, some of which will have depreciated since the last valuation, rather than take new assets on their own. These tactical manoeuvrings can affect the project and are often in the unknown.

As the key political agenda item of 'patient choice' moves into place, then the PCT and even individual GP practices, which can hold indicative budgets, become part of the route to funding and the director of services. Ultimately, some of the work may go to private hospitals that have procedurally engineered basic procedures to reduce costs. Within this environment, trust finance directors tend to be extremely aware of current issues, see potential sources of funding, and be creative in designation of the trust's actions to achieve maximum funding. There is of course monitoring of this, both by accountants and by the Audit Office and the Healthcare Commission. These are all political stakeholders in the trust's success. Trusts seek as little interference in their management as possible in order to maintain a strategic course of action. As the political environment changes and internal events take shape, however, it is difficult to avoid problems, and this creates tensions, which are in the unknown.

There are advantages in having large hospitals that can provide the full gamut of general services and even some specialist services of national significance. Such hospitals have economics of management and of other facilities for staff and patients. Large trusts are significant players in the political environment both for the delivery of political agendas and for them to place political pressure on authorities. Against this, however, is the problem of managing such a large organisation and in making the care friendly to patients. Another move is for trusts to achieve foundation status. This makes a foundation trust into an individual public-benefit corporation, which, although not privatised, gives the trust significantly more organisational independence. This is controversial, as it prioritises these trusts against others and has a background of privatisation.

Professional/operational domain

Modern medicine in a hospital involves a large number of specialists using high-technology equipment and a vast array of drugs to make people well and alleviate suffering. The nature of this means that patients come to the hospital on an outpatient or inpatient basis. This flow of people through the hospital (inpatient spells) has many aspects, including how they arrive and whether they are elective (admitted from waiting lists), which could be day-care, or non-elective (admitted as urgent or emergency). All require different facilities during their 'spell', for example accommodation, nursing, diagnostics, prescribing, food and direct medical treatment, such as surgery. As well as the patient, there are family members and other companions who require attention in order to support the patient. A key aspect of this is the production, storage and retrieval of information. As the patient moves through the system and is dealt with by different people, the patient's medical information must travel with them to enable successful treatment and efficient delivery. Data capture is also required for management and audit and has given this administrative area a growing importance particularly, as the use of information technology (IT) is regarded as a way of reducing costs of delivery of service.

The number of different professional parties in the hospital is large. The central figure in this is the consultant under whose care a patient is designated. The authority and power of the consultant are legendary, although they are also skilled and dedicated individuals. Such an expensive and sought-after resource needs to be used effectively, and their support teams become critical in doing this. The medical team under the consultant includes registrars and other junior doctors, who require training and the opportunity afforded to their careers by the consultant. The consultant's degree of discretion in medical matters and hospital organisational matters is now being challenged. With regard to the former, consultants are now subject to clinical governance and open to clinical audit and the publishing of success statistics (as promoted by NICE). With regard to the latter, consultants' contractual arrangements are being changed to make them employees, thus exposing them to having their time managed and curtailing their ability to undertake private work. This is clearly a change instigated in the political domain, which affects the operational domain both in efficiency and effectiveness. Some consultants' private work may be undertaken in private hospitals, which are in competition for some procedures within the new 'patient choice' agenda and this can cause operational tensions in the trust.

Within the hospital, there is a degree of professional competition between medical disciplines. This may not be apparent in normal patient care, but as departments vie for resources and individuals wish to work in renowned departments because departmental status can enhance careers, then this competition may appear. Individual professional status comes from peer-group renown, either from publishing research or speaking about new medical procedures. This may induce a conflict with the hospital management, which is interested in success rates, speed and quantity of delivery, patient satisfaction and lower operating costs. The dilemma for management is that the establishment of a specialist area, which can then command extra resources and give the hospital renown, can strategically deliver an enhanced position. This more active research environment makes the hospital attractive to the best staff in order to recruit and retain them. This is easier in hospitals that are attached to a university and involved in teaching and research. There is a tension between medical specialisms under a managerial system that is restricting resources and also trying to implement managerial operational controls while at the same time advancing medical practice.

Medical support such as radiography and pathology has its own departments and also often its own consultants. In the changing technological world of medicine, these disciplines experience and are stewards of much of this change. Their status is increasing, and so their authority in the hospital is increasing. Nursing is becoming a graduate profession and is changing its role from patient care to patient management and ward management. There are now consultant nurses and specialist nurses who have some medical authority and are allowed to prescribe and undertake procedures (e.g. surgical care practitioners). There is tension between this and the medical world, which wishes to retain its status in society (Macdonald, 1995). There is also a tension within the nursing world, where the practice of nursing is being taken over by theory. The less pleasant tasks of patient care are being devalued, and patient satisfaction is being diminished. Ward housekeeping and patient-care duties are also carried out by healthcare assistants (nurse auxiliaries). The supervision of ward cleaning was taken away from nurses but was then returned because of problems with infection control, particularly with methicillin-resistant *Staphylococcus aureus* (MRSA).

The care team also involves people in social work and occupational health roles for both the patient and the family of the patient. The number of allied health professionals is growing, and the management of them and the extra information they are bringing to decisions adds to the complication. It is difficult to see the whole picture of the patient and their condition among the disparate practitioners who work on different parts of the patient.

The role of facilities and estate management is also changing, with some directors of estates commanding board positions. These are also the managers of new developments and so have considerable influence in the practical and functional aspects of buildings and an awareness of the internal operational and political problems that are exposed during building. This is a developing role, which means that it is different in different hospitals. The role of the building, however, is not only functional, although the financial pressure drags it this way. It is realised that the physical infrastructure is expensive and critical in both the successful operation of the hospital and in the efficient use of resources. Hospitals are locations that require a positive identity for communities and hospital employees. A positive environment, both physically and psychologically, is required to assist healthcare. It is difficult to justify the resource for this, however, against the immediacy of funding, the functional needs of supporting people in their illness, and delivering cures.

Managerial domain

Decisions about how the hospital operates are made in the managerial domain. This world has a high degree of rationality, both in the forms of information it produces and in how it produces this information. These managers need to manage the boundary between the professional/operational domain and the political domain as well as the divisions within the professional/operation domain.

Strategically, the political negotiations with the SHA and the various PCTs determine the direction in which the hospital is allowed to go in modifying its services and building new facilities. The key to this is the creation of a strategic outline case (SOC) and outline business case, which ultimately will be made into a full business case (FBC). This SOC involves the taking of an idea and making it fit the prevailing needs, while also making it fit in a way that is financially viable and practically doable. Indeed, it can be the case that there are agendas for projects that are not explicit in the plans. For example, if the hospital needs its carpark refurbished but it cannot afford to do this itself, then it demonstrates a need for a carpark as part of a delivery of medical outputs, which it can apply for funding for. This hidden objective is known to the senior managers but may not be apparent elsewhere. Such creativity requires a high degree of awareness of which priority areas will be funded, so that the case can be written to include this.

There is a question about how much discretion hospital management has in its strategic decisions or whether, in reality, all agendas are set outside, which management simply implements. Thus, decisions about what service to commission are made at the SHA level with the PCT, overall priorities are set by the DH centrally, and doctors only do what they will do. This question is played out in the business case, with the hospital trying to sell its strategy against the agendas set outside and also selling its strategy internally against the competing demands. Opposition or misalignment is then discovered, and the case may fail or be modified. This manoeuvring creates strange and messy projects, where the objectives are a complex mixture of external demands and internal desires.

Tactically, hospital management becomes a management of information, which involves the collection and presentation of statistics. In many ways, the meeting of targets that have been set by the Health-care Commission (e.g. waiting time for cancer treatment) becomes the primary management objective. In 2005, there were eight key targets and 32 balanced scorecard performance indicators to be managed; the list is shown in Table 7.1 for 2004–05. These allowed the external agencies to monitor individual hospitals. The balanced scorecard approach has been promoted within the public sector as a way of balancing financial assessment against customer service, business process operation and longer-term learning and growth (Kaplan & Norton, 2000). These targets or healthcare standards change annually and are expected to prioritise the direction of action by the trusts. As in the strategic management arena, they have political as well as operational consequences. The underperformance in an area or even the loss of a star rating overall becomes an issue of high strategic importance – and even more so when foundation status has been achieved. In 2004–05 hospitals were losing stars mainly because of inadequacy of financial performance, and this induced political pressure to replace the senior management of such hospitals. In response, hospital management undertook cost-cutting measures, including staff redundancies, which impacted on the patient care and the local political arena.

The 12 royal colleges (e.g. surgeons, physicians, pathologists) represent specialist consultants and periodically make pronouncements about maintaining standards on medical procedures, staffing levels and appropriate medical environments. If there are problems in a hospital, as reported by a college member, then the college may produce a report and make recommendations for action by the hospital. The royal colleges have no formal role but represent stakeholder power, such that if a hospital fails to respond adequately the colleges can influence the managerial domain considerably by threatening to access the political domain.

Operational management involves using the minimum of resources to undertake tasks and reducing the overhead of this operation. There is a political belief that there is much waste in the system and that the service could be maintained at a lower cost if this waste was eliminated. This requires process-mapping and process control. Patients become the material to be processed. Thus, the managerial world talks about 'patient pathways', i.e. the sequence from entry to exit within the system. This involves a number of inpatient stays, all of which must be measured and compiled into the statistics. This information management is different from that associated with patient medical

Table 7.1 Key targets and measures for 2004–05.

(a) Eight key targets
1   12-hour waits for emergency admission via A&E post-decision to admit
2   All cancers: 2-week wait
3   Elective patients waiting longer than standard
4   Financial management
5   Hospital cleanliness
6   Outpatient and elective (inpatient and day-case) booking
7   Outpatients waiting longer than standard
8   Total time in A&E: 4 hours or less

(b) 32 balanced scorecard measures
1   A&E emergency admission waits (4 hours)
2   Better hospital food
3   Breast cancer: 1 month diagnosis to treatment
4   Breast cancer: 2 months GP urgent referral to treatment
5   Cancelled operations
6   Child protection
7   Clinical risk management
8   Composite of participation in audits
9   Data quality on ethnic group
10 Deaths following heart bypass operation
11 Deaths following selected non-elective surgical procedures
12 Delayed transfers of care
13 Emergency readmission following discharge (adults)
14 Emergency readmission following discharge for fractured hip
15 Indicator on stroke care
16 Information governance
17 MRSA
18 Outpatient and A&E patient surveys: access and waiting
19 Outpatient and A&E patient surveys: better information, more choice
20 Outpatient and A&E patient surveys: building closer relationships
21 Outpatient and A&E patient surveys: clean, comfortable, friendly
    place to be
22 Outpatient and A&E patient surveys: safe, high quality, coordinated care
23 Patient complaints
24 Patients waiting longer than standard for revascularisation
25 6-month inpatient waits
26 Staff opinion survey: health, safety and incidents
27 Staff opinion survey: human resource management
28 Staff opinion survey: staff attitudes
29 13-week outpatient waits
30 Thrombolysis: composite of 60-minute call to needle time and 30-minute
    door to needle time
31 Waiting times for rapid access chest pain clinic
32 Workforce indicator

A&E, accident and emergency; GP, general practitioner; MRSA, methicillin-resistant
Staphylococcus aureus.

care, but they are managed together within the IT system. A large trust may have over 6,000 employees treating over 500,000 patients a year. The management of these is expected to be transparent and consultative, which requires considerable time for decision-making and the production of large quantities of information and statistics.

This operational management undertaking is complex and open to heavy scrutiny when things go wrong. In a sense, a hospital is a queuing system, with patients waiting to be processed through admission, assessment, diagnostics, treatment, recovery and discharge. Elective patients arrive from a waiting list; non-elective patients are urgent or emergency admissions. Problems of availability of beds, equipment and consultants can seriously disrupt the flow, such that elective treatment may be postponed because of the arrival of urgent non-elective cases. Increasing the throughput of electives may degrade the service and anger staff. Bed management becomes an everyday activity and a role in its own right, whereby the timely discharge of patients is enabled in order to release beds. Although certain wards are designated to certain medical specialisms, patients may have to be placed in wards not related to their condition. Having high bed occupancy reduces overheads and makes the best use of capital, but this can produce operational problems.

Structure

Hospitals are large organisations and therefore require considerable organisational structure to be manageable. This tends to be hierarchical, and trusts have numerous organigrams to demonstrate reporting and decision-making committee structures. This reflects the high degree of bureaucratic accountability that is expected in the managerial domain and the degree of consultation required across the divisions. The processes are highly rational, both in a logical sense and in a sense that they are fixed and are expected to be followed. The information in Table 7.2 gives a snapshot of the management of a particular trust. Although there are many differences in detail between trusts, all have a similar overall arrangement.

The main board of a trust sits beside an executive management team. The main board has more strategic and corporate governance agendas, whereas the executive management team deals with operations and brings forth strategic plans to the main board. Trusts are headed by a chief executive, and the executive management team reflects the divisions of concern, i.e. finance, medical, nursing, operations and human

Table 7.2   Snapshot of Buckinghamshire Hospitals NHS Trust, 2004.

Trust board

Chair
   Professor David Croisdale-Appleby
Non-executive directors
   Mr Mike Bellamy
   Mr Brian Chapman
   Mrs Christine Sivers
   Mrs Eileen Springford
   Ms Pier Thomas
Executive directors
   Chief Executive: Ms Ruth Harrison
   Medical Director: Dr Andrew Kirk
   Director of Nursing: Mrs Maureen Davies
   Director of Finance, Performance and Information: Mr Gordon Greenshields
   Director of Operations (Wycombe): Mr Malcolm Newton
   Director of Operations (Stoke Mandeville): Mr John Blakesley
   Director of Human Resources: Mrs Sandra Hatton
   Director of Strategy and Communications: Miss Sheryl Knight
   Director of Facilities and Estates (including PFI): Mr John Summers
Clinical directorates
   Accident and emergency
   Anaesthetics
   Medicine for older people
   General medicine
   General surgery
   Head and neck
   Obstetrics and gynaecology
   Paediatrics
   Pathology
   Plastics and burns
   Radiology
   Specialist medicine
   Spinal injuries
   Trauma and orthopaedics

This trust involves three hospital sites: Amersham, High Wycombe and Stoke Mandeville. The trust serves residents in Buckinghamshire, Thame (Oxfordshire), Tring (Hertfordshire) and Leighton Buzzard (Bedfordshire), a combined population of 500,000. It serves a much larger population of 1.5m for burns and plastic services and 14m for spinal injuries, being the national spinal injuries centre.

Three primary care trusts (PCTs) are responsible for commissioning healthcare for this population: Vale of Aylesbury PCT, Chiltern and South Bucks PCT and Wycombe PCT. Bedfordshire and Hertfordshire also send patients to these hospitals. In 2004 the trust had 4,700 staff and an annual budget of £185m. It has assets of about £190m and spends £10m annually on buildings and equipment.

Under the government's private finance initiative (PFI) initiative, Stoke Mandeville has a £40m development for 10 new wards comprising 222 beds, two new day-care operating theatres, an endoscopy suite and a new main entrance, with restaurant and retail facilities.

Enterprise Healthcare Ltd, which leads the private-sector consortium, comprises Alfred McAlpine Project Investments Limited (construction), Sodexho Investment Services Limited (facilities management) and HSBC Infrastructure Fund Management Limited (equity investor).

Source: Buckinghamshire Hospitals NHS Trust (2005) Annual Report 2003–2004. Aylesbury: Buckinghamshire Hospitals NHS Trust.

resources. The main board team may also include: informatics, clinical governance, policy and planning, special projects, research institutes, and university and facilities directors, depending on the circumstances of the trust. Large trusts may be split into divisions that reflect medical specialisms or different hospital sites, and each of these will have a director on the executive management team. These divisions are run as subunits and have structures that match the main executive team structure. The trust's main board includes some of these executive team members plus a chair and non-executive directors. There is normally a match of numbers between the executive and the non-executive teams. This creates a political balance; however, personalities are also important in board decisions. Non-executives assist the trust by providing external probity to matters such as chairing senior appointments committees and remuneration committees. However, they introduce external stakeholder issues and a dynamic that may be at odds with the executive team. As well as these voting members, a number of non-voting members from the executive management team will sit on the main board.

7.3   EXPERIENCE OF BUILDING: FROM UNKNOWNS AND CONTRADICTIONS TO MEANS AND ENDS

The diagram in Figure 7.3 attempts to summarise the complexity of the knowledge and processes within an NHS trust. It shows two contributing domains – the situation itself and the thinking within the situation – which determine the third domain of action. Of course, action feeds back to the situation and thinking domains. This world is full of gaps and contradictions, which induce action in the means-and-ends dimension and so determine a trust's response to building.

Wants–resources dilemma

The health world is full of wants and potential opportunities to provide for them; however, there are both absolute and local resource limitations that stop these wants being fulfilled. Clearly this is the case within illness-curing itself, but it is also exhibited by managerial problems because of staff shortages and individuals' concerns about their careers. In a sense, these extensive wants are the cause of the policy problem. The difficulty is in distinguishing between needs and wants. Needs are critical and established by a norm. Beyond needs are

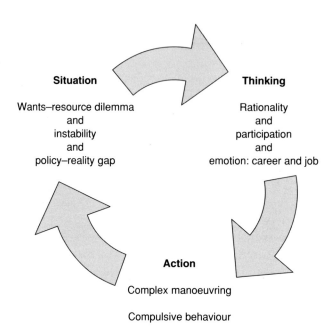

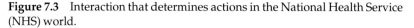

Figure 7.3   Interaction that determines actions in the National Health Service (NHS) world.

wants – extras that can be provided but whose fulfilment does not add much generally. From an individual healthcare worker's or patient's point of view, their needs and wants are essential and their lack of provision produces a feeling of loss and resentment.

Policy–reality gap

The high amount of resources that are employed and the high wants mean that a politically structured approach through policy is utilised. This means that it is high on political agendas as it impacts directly and emotionally on electorates. The promises set up by this political policy-making against the rationing of resources create a gap between the rhetoric of policy and the reality of possible implementation. Of course, this gap is handed to the health-delivery managers, who learn to cope with balancing the reality with the policy. In addition, the health world is dominated by potential new drugs and procedures, which also have a gap with reality. These political challenges into the rationality of the medical way of thinking and these gaps add to the emotional nature of the world of sick people.

Instability

The world of health, politically, managerially and operationally, is changing, and this dynamic creates further emotions within individuals and organisations. Change is endemic because of the belief that there is always a better way of coping with the needs–resources dilemma. This induces change in policy at the top political level, as new ideas are generated to do things better. It induces further managerial change, which is coping with the change in policy and also seeking better ways of doing things. Issues with change include many people not understanding it or the implications of it. Thus, the system is always learning to cope with different perspectives and filling in the gaps where the change has not been thought through or has been implemented unsuccessfully. In this change, there are many stakeholders and agencies whose positions may not be known but subsequently can influence outcomes.

The world of the acute trust involves many people who are in the system temporarily either because they are agency staff or because teams are reshuffled. This means that teamwork may not be developed, which can induce further instability. As patients are cared for by different people as a result of different specialisms and patients' extended time in the system, procedures of record-keeping are required in order to allow the care to move from one person to another and still be accomplished. This temporal and personal fragmentation is a stress on the system and a cause of discomfort in patients. Integration to some extent occurs through rationality, i.e. by procedures and learned ways of doing things. However, it has been identified that practices across the country differ widely, and this affects the quality and efficiency of service. The NHS works through the commitments of staff to make it work. They overcome gaps by asking questions and putting extra work into ensuring the gap is covered. This in itself induces instability, as different people do things differently, but the rationality of the whole cannot properly acknowledge this and so it is an organisational contradiction.

Rationality

The world of hospitals is a world of rationality in three ways. First, powerful figures have a scientific belief in the power of rationality, as this is a medical world. Thus, clever individuals can think through situations and order them to effect better action in ideal circumstances. Second, the policy-controlled world is based on a belief in positivistic planning, where grand plans are deliberated over and then implemented and the change controlled against the plan. Third, the temporal and personal fragmentation involved in caring for patients requires procedures. This rationality is compounded by the need for high accountability and transparency, where everything is written down and made available publicly. Rationality becomes a protection from the problems of the system. This is exaggerated now by the need to manage through performance indicators, all of which require the collection and processing of statistical data.

Everything is procedurised in detail, which is both a limiter and an enabler to effective action. Standardised care is more efficient and prevents errors. Where organisational procedures do not work, however, there have to be procedures in place to deal with this. It may be that the task is obscured by the complexity and comprehensiveness of the procedures. Indeed, the quantity of procedures makes it difficult to understand them all in detail, which is made worse by the instability that is changing them. The whole is required to work together, vertically and horizontally, in order to make a patient better, but this cannot happen completely rationally. Thus, there is a need for highly participative decision-making.

Participation

There is a belief in consultation and participatory decision-making involving all stakeholders. This is required because of the complexity of the situation and the high degree of differentiation of activities that need to be brought together in order to make things work. It is also necessary because it is a world of people who care. Thus, there is a contradiction between the belief in rationality and the belief in people who are, to some extent, non-rational. The complexity of the situation means that no one has a complete understanding, and so partial understandings are used by stakeholders in decision-making. Solutions that come out of such complex task and process interactions are, by their nature, inadequate. This introduces the emotional world; thus, decision-making becomes the forum for individuals to work through their emotional needs. There are attempts to procedurise decision-making in order to make it appear rational, even when there are real and hidden conflicts. Participation has the tendency to deliver solutions that are fragmented and involve the least unacceptable outcomes. The process itself is emotional, and the outcomes, being inadequate, are also emotional.

Emotion

Emotion is important to the way in which people think in the health service because of the nature of the job and because of individuals' positions and careers within it. There is a contradiction between this way of thinking and rational thinking. The inadequacies of the latter create emotion, which adds to an already emotional world of sickness. Being sick is emotional, and being a relative of a sick person is emotional. Sickness evokes the emotions of helplessness and inadequacy. Thus, sickness induces a feeling of dependency that someone can or, indeed, should help. There is also an emotion that sickness is someone else's fault, which can reveal itself as anger. This is made worse by a society that now expects instant relief (the 'magic pill') and believes that money can make anyone better (Furedi, 2005). To work in a world of sickness, people need to learn to contain their own and others' emotions (Obholzer, 1999). This can involve a complete range of reactions in health workers, from detaching themselves completely from the sick (e.g. at a policy level or via medical procedures) to empathising completely with the sick (e.g. counselling or religious ministering). These differences display themselves in the different professions and in the arrangements of decision-making.

Healthcare is also a world of rationed resources. Thus, in addition to the emotion of someone being sick, it may not be possible to help them because of a shortage of resources. A further emotion comes from the competition for resources that induces conflict between different professional groups and departments. In a participatory world, this is often displayed during decision-making meetings. This competition for resources fuels a degree of resentment of the situation but also the development of ways of manoeuvring in situations.

The health service is also emotional because of the difficulties involved in evaluating new technologies and techniques, which call for resources. In order to obtain these resources, it is necessary to focus on the positive features of the technologies and techniques. If these do not deliver or negative features are significant, there may be calls to abandon these techniques. This interacts with individual career aspirations.

There are careers to be made and status to be achieved through medical success and managerial determination. People are interested in, and get quick advancement through, working on new techniques and dramatic changes (Dopson & Mark, 2003). Seeing others obtaining resources that may help their careers may induce negative emotions.

As status and position are dependent on patronage, personal alignment is critical. Clinicians can create fiefdoms with strong boundaries, which have been identified as indicators of deeper organisational problems (Protopsaltis *et al.*, 2002). These internal boundaries make wider decision-making difficult and lead to further manoeuvring. People's positions can be threatened by changes: as some people are advancing their positions, others may see theirs being eroded, and so battles ensue.

This is made worse by the wider system, such as politicians and local media, using these difficult and emotional situations for their own purposes. In treatment situations, difficult decisions have to be made, and non-favoured outcomes can transpire. Thus, as decisions are made, there is a degree of defensiveness that can make the decision more difficult. The potential influence of these wider players is used to promote particular courses of action, which further adds to the manoeuvring.

Action

The hospital as a situation is so complex that no one understands, or is in full control of, it all (Sweeney & Griffiths, 2002). Everyone sees it differently, understands it differently and gets something different out of it. The emotion from this and the emotion of illness generate compulsive behaviour. In this, it is easier for people to look inwards at their jobs in order to maintain conditions rather than to look at the operation of the wider system. In this, they seek to make something meaningful for themselves and set their own performance criteria. Thus, decisions are made by sets of individuals, negotiating using their power and by managerial juggling. The rational organisation is confused by this hidden manoeuvring and the hidden agendas in objectives. The NHS works through the commitments of staff to overcome gaps and contradictions through extra work. This commitment to overcome system inadequacies empowers staff members, and they feel justified in their manoeuvring but resentful of needing to do it.

In the current environment of targets, there is a tendency to manoeuvre in order to create a good impression. This makes contorted decision-making but also directs thinking to short-term benefits. The nature of politics means that staving off attack is a daily occurrence. Many managerial staff see the best form of defence as to concentrate on the operational world rather than the strategic world. The emotion induced by this creates a distance from the difficult issues of daily management, and thus the world becomes one of managing assets.

Policy is used to support agendas; it is also used in a creative way to seek resources for special situations.

7.4   MEANS AND ENDS OF BUILDING

The world of hospitals is a world of apparent rationality. Powerful figures have a scientific belief in the power of figures, positivistic planning and procedures. There is a need for high accountability and transparency. This is exaggerated now by the need to manage through performance indicators. There is a belief in consultation and participatory decision-making. However, the rational whole is confused by hidden manoeuvring and agendas hidden in other objectives. It is also a world of rationed resources, and therefore there is competition for resources and a degree of resentment of the situation. There are careers to be made and status to be achieved through both medical success and managerial determination. This creates an emotional world on top of an already emotional world of illness. The world of health, politically, managerially and operationally, is changing; this dynamic creates further emotions within individuals and organisations. The known and unknown means and ends are presented in Figure 7.4.

When an opportunity for development occurs, the emotion that was suppressed from the needs–resources dilemma comes out and creates high demands. The uncertainty induced by the opportunity of building and the potential improvement in the organisation leads to high emotion. These desires are beyond what the system can meet, which induces compulsions for wanting more from the same money and manoeuvring in order to achieve this. Most people have a genuine desire to make things happen and make things better; however, their intentions can get confused between improving the service and improving their personal conditions. This is rationalised by the importance of saving lives and making people better.

Construction is seen by many NHS acute trust staff as a problem to be fought. It is believed that construction should do more for less, that it is profiteering from health, and that it is refusing to do a good job. There is a general distrust of the private sector, which creates a siege mentality, so that relationships with external stakeholders can be poor. If staff have not experienced negative relationships with the construction industry themselves, they will know of and be able to talk about others' negative experiences. The construction industry has

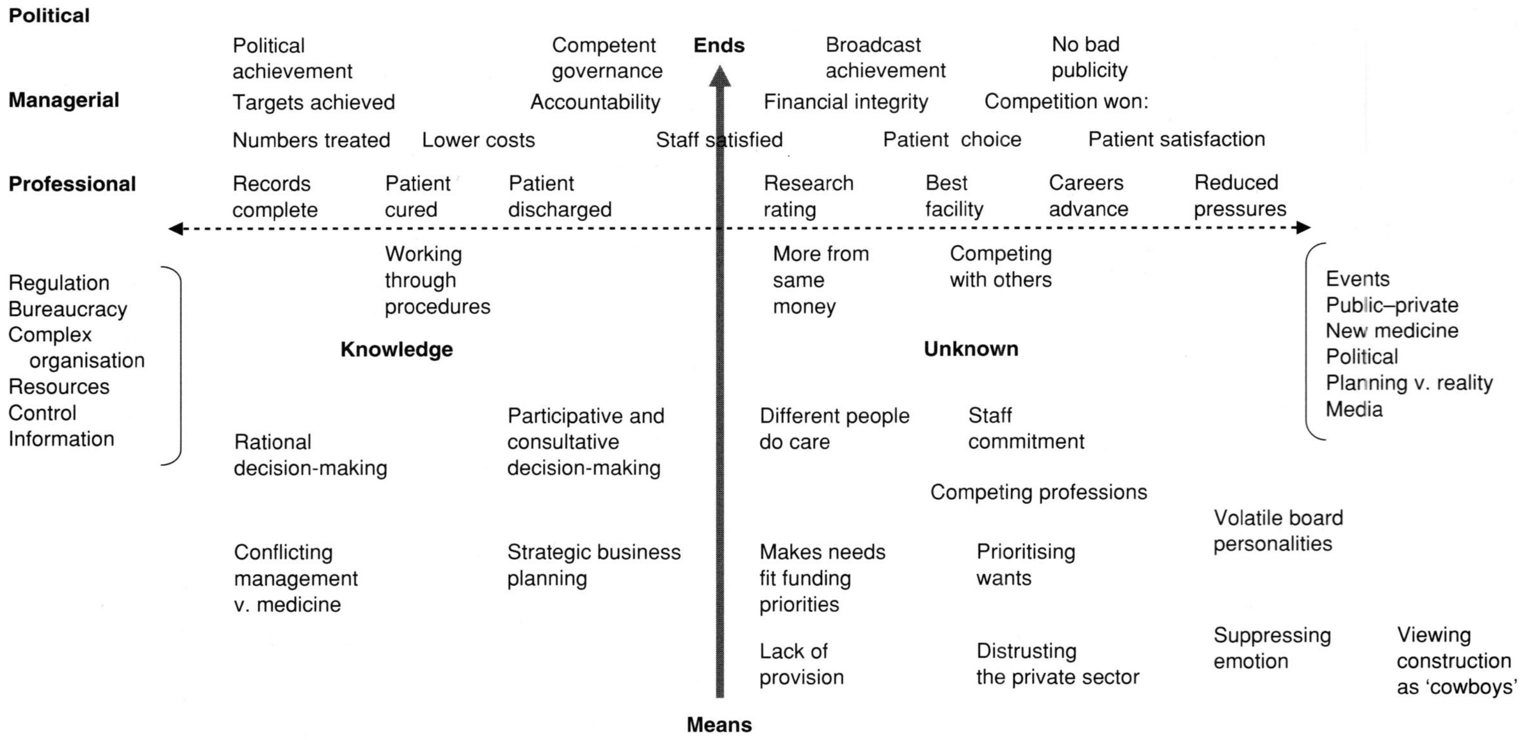

Figure 7.4 National Health Service (NHS) acute trusts means and ends.

a history of being seen as the generator of conflict. Different stakeholders seek out different members or disciplines of the construction team with whom to align. The industry fragmentation allows this to happen, and so the compulsive behaviour from the health service is transferred to the industry. Architects are sought by those wanting special quality, quantity surveyors are sought by those wanting to reduce costs, engineers are sought by technology disciplines, and instant buildings are sought by those interested in time.

There has been a history of procedurised approaches to capital funding and construction (e.g. Capricode, capital procurement manual). There is a history of procedurised design, each with its own label, e.g. Nucleus Hospital, Patient Journey Model, Better By Design, NHS Design Quality Portfolio Technical and User Criteria, PFI Design Development Protocol and Model Design Quality Specification. There is also a history of how to manoeuvre using these procedures. Procure 21 was a new means of procurement developed in 2000 as an attempt to avoid conflict rationally in NHS trusts' problematic relationships with the construction industry. Trusts, however, have not been completely supportive of this and have manoeuvred to not use it or to use it only partially. If the internal leadership recognises the problem, then such procedurised moves can help by getting both NHS and the industry staff to understand risks by, for example, using the NHS ProCure21 design and risk tool (DART: www.nhs-procure21.gov.uk) and engaging staff in open informed decision-making.

There are questions that rationality and participative decision-making cannot answer. Thus, spending money on a building takes away resources from new medical technology or more staff, but new medical technology requires buildings and staff complain about the poor state of buildings. This has induced a piecemeal approach to development, whereby minimum extensions are provided and hospital sites become patchworks of styles and conditions. The problem is seen as the buildings and the construction industry rather than a problem of decision-making.

7.5   KEY ISSUES

In summary, the key points of the health sector are thus:

❏ The NHS world is so complex that there is no single understanding of it.

❏ It is in constant change.

❏ It is rational and expects highly rational processes.

❏ It requires accountable and participative decision-making.

❏ Although rational, its decision-making exhibits considerable emotion.

❏ Shortage of resources induces departmental manoeuvring, which obscures project purposes.

❏ Decision-making is difficult in a tightly resourced and changing world.

❏ There is a dependency on the industry but a belief in having power over it.

❏ There is a distrust of private-sector business.

❏ There is a tendency to believe that you can always get more.

❏ NHS processes may be manoeuvred around for particular purposes.

❏ Project management requires understanding and being open but firm.

❏ Compulsions in the healthcare team may be aligned with individuals in the construction team.

REFERENCES

Dopson, S. and Mark, A. (eds) (2003) *Leading Health Care Organisations*. Basingstoke: Palgrave.

Emmerson, C., Frayne, C. and Goodman, A. (2002) How much would it cost to increase UK health spending to the European average? Briefing note 21. London: Institute for Fiscal Studies.

Furedi, F. (2005) Our unhealthy obsession with sickness. Presented at the Institute of Ideas Conference, London, 12 February 2005. www.spiked-online.com/articles/0000000ca958.htm. Accessed 28 March 2006.

HM Treasury (2004) Spending Review 2004. www.hm-treasury.gov.uk/spending_review/spend_sr04/spend_sr04_index.cfm. Accessed 24 March 2006.

Kaplan, R. and Norton, D. (2000) *The Strategy-focused Organization: How Balanced Scorecard Companies Thrive in the New Business Environment*. Boston, MA: Harvard Business School Press.

Lefty, M. (2005) Kill or cure? *Building*, **8 July**, 45–47.

Macdonald, K. (1995) *The Sociology of the Professions*. London: Sage.

NAO and Audit Commission (2005) Financial Management in the NHS. Report HC 60-I 2005-2006. London: National Audit Office and Audit Commission.

O'Neill, O. (2002) *A Question of Trust*. Cambridge: Cambridge University Press.

Obholzer, A. (1999) Managing the unconscious at work. In *Group Relations, Management and Organisation*, eds R. French and R. Vince. New York: Oxford University Press.

OHE (2000) *The Economics of Health Care*. London: Office of Health Economics.

Protopsaltis, G., Fulop, N., Meara, R. and Edwards, N. (2002) *Turning Around Failing Hospitals*. London: NHS Confederation.

Sweeney, K. and Griffiths, F. (eds) (2002) *Complexity and Healthcare: An Introduction*. Abingdon: Radcliffe Medical Press.

Wynne, A. (2002) PFI and the Public Sector Comparator: Are Comparisons Really Objective? London: Association of Chartered Certified Accountants. www.accaglobal.com/publications/accountingandbusiness/380769. Accessed September 2005.

RESOURCES

Gatrell, J. and White, T. (2003) *The Specialist Register Handbook*, 2nd edn. Abingdon: Radcliffe Medical Press.

Prowle, M.J. (2001) *The Changing Public Sector: A Practical Management Guide*. Aldershot: Gower.

Shortell, S. and Kaluzny, A. (2000) *Health Care Management: Organization Design and Behavior*. New York: Delmar.

British Medical Association: www.bma.org.uk

Department of Health: www.dh.gov.uk

Healthcare Commission: www.healthcarecommission.org.uk. Independent inspection body for both the NHS and independent healthcare.

Health Facilities Management Association: www.hefma.org.uk

National Institute for Health and Clinical Excellence (NICE): www.nice.org.uk. Independent organisation responsible for providing national guidance on the promotion of good health and prevention and treatment of ill health.

NHS Confederation: www.nhsconfed.org. Brings together organisations that make up the modern NHS across the UK.

NHS Gateway: www.nhs.uk. Information portal to the NHS.

NHS Institute for Innovation and Improvement: www.institute.nhs.uk. Provides the focus for new ideas, technologies and practices to improve services to patients, users and the public.

NHS Purchasing and Supply Agency: www.pasa.nhs.uk

Patients Association: www.patients-association.org.uk

Royal College of Nursing: www.rcn.org.uk

NB: The NHS changes regularly, and with this change web resources are deleted and new resources developed.

8 Governments as Clients

8.1 INTRODUCTION

In a democratic post-industrial society such as Britain, government is the organisational face of the nation, both as its representative and as its organiser of social enterprise. Governments are elected, and seek to be re-elected, in order to serve society within a political and economic system; thus, governments have, and represent, power. Election involves the presentation of ideas against other people's ideas for actions that improve the country's international role, the effective delivery of services and the upholding of the law. This party battle requires any positive outcomes to be attributable to the government and any problems to not be the fault of the government. The nature of a multiparty democracy is that the opposition needs to have a separate identity and a separate approach to action, which they can promote to the electorate as better than that of the current government. Any party's policies are heavily dependent on its history, in particular the successes and failures of their previous administrations.

Governments create policies, which are authorised by a vote in Parliament in which they have a majority. These policies, set within ministries, direct the strategic course of action that will be undertaken. Each ministry has a civil-service department that administers that ministry's actions. Governments believe that the civil service is there to undertake their agendas; however, the civil service believes that it needs to be independent and offer a continuity of transparent accountable action across different governments. Governments are composed of ambitious people who seek career advancement and who may not agree with others in the government; thus, there is a degree of competition between ministries for resources and authority. This ambition and competition are also present in the civil service, and this aspect of partisanship towards government and other ministries is managed by a set of bureaucratic procedures and hierarchical decision-making. These procedures provide an essential degree of accountability and a barrier to excesses of government.

There are many construction clients within the government sector. We cannot discuss them all, but we cover three, with the anticipation that these will provide a representative insight into how government clients operate. The three sectors that we have chosen are: Defence Estates, the Highways Agency and Her Majesty's Prison Service (HMPS), which are located within the Ministry of Defence (MOD), the Department for Transport (DfT) and the Home Office. We will use domain theory, as presented in Chapter 3, to understand the way in which each of these clients delivers its objectives to, and receives resources from, its environment. Domain theory distinguishes both a different purpose and a different set of achievements at political, managerial and professional/operational levels. Thus, any issue or event will have an aspect that will be seen differently through each domain.

8.2 THE POLITICAL DOMAIN: SERVICE IN A POLITICAL ENVIRONMENT

All government sectors share the overarching political context of the current government. This wider political position is set by a mixture of ideology and the history of events. We discuss these with reference to the delivery of public services and, in particular, a government agenda towards construction in general. Following this, we discuss the specific service that each of the sectors undertakes within the government context. The departments chosen are quite different, have a different authority within government and have had quite a different history. This gives each a character that very much reflects this authority and the difficulties of delivering their purpose.

One of the big agendas of government is managing the economy in order to provide services that are seen to be effective. The expenditure in the different government functions is shown in Table 8.1.

Governments and opposition create strategic agendas and policies around this management of the economy. Since the 1980s, these include an agenda that the public service could have a more efficient operation, that services could be delivered better and at a lower cost for the public, and that some of this could be achieved better

Table 8.1   Government expenditure by function as percentage of gross domestic product (GDP).

Estimated outturn	2004–05 (£bn)	% of GDP
Social protection	138	28
Health	81	17
Education	63	13
Law and protective services	29	6
Defence	27	6
Debt interest	25	5
Other health and personal social services	22	5
Industry and agriculture	20	4
Housing and environment	17	3
Transport	16	3
Other	49	10
Total	488	100

Source: Office for National Statistics (2005), p. 365.

by private-sector companies. Asset management is a key part of this agenda. Asset management includes selling surplus assets in order to free up resources for new investment, and transferring ownership of assets to the private sector in order to secure access to new funding and skills or to transfer risk. In addition, there is a drive to identify and capitalise on hidden assets and to increase value for money from retained assets. There has been considerable change in the operation of public-service procurement since the 1980s. Although much of this is operating in ministries, the Office of Government Commerce (OGC), which is part of the Treasury, directs and coordinates many of the strategic changes. In addition, the delivery of government services, in particular the effective expenditure of public funds, is reviewed by the National Audit Office (NAO), which is an independent body outside the government with powers to report directly to Parliament. There are other audit bodies that relate to the devolved governments and assemblies within the UK. Together, these bodies oversee the economy, reporting on the efficiency and effectiveness of government, ministries and departmental services.

Construction is only one of the services that the government procures. The scale of government, with an asset base of about £220bn in 2005, means that it is the major customer for most services and products in the country, even with the changes away from the public sector. This scale provides the government with problems because of managing the quantity of any one aspect and having a consistency in their approach. There is a belief that many aspects of this procurement are inefficient and involve waste. The OGC has a general remit to deliver improved procurement efficiency and to be able to demonstrate value for money. It undertakes this through various promotional teams that support departments across government and through measuring and reporting on efficiency gains. It has created a gateway process to examine a procurement programme at five critical stages during the lifecycle of any project – three before contract and two looking at service implementation and confirmation of benefits (see www.ogc.gov.uk).

The Kelly Report (OGC, 2003a) looked at demand and capacity in the public-sector supplier marketplace in a bid to balance the strategic benefits to both government and suppliers. The intention was to make the public sector better at planning its procurement and communicating its needs to industry while also ensuring that there were no adverse consequences of overdependence on a single source (OGC, 2004). Government departments are now required to reduce red tape in procurement, improve the attractiveness of the government marketplace to all suppliers, stimulate competitive responses from suppliers, and investigate the management of supply chains.

Construction is one aspect that is being targeted because of a continuance of poorly delivered iconic projects and the overall statistics from the NAO demonstrating, in general, poor value for money. Indeed, the extensive number of reports on the construction industry (Murray & Langford, 2003) is derived particularly from the failure of government procurement, and this has been the case for many decades. Earlier studies tried to sort out construction through further bureaucratic procedures, onerous contracts and standardised briefs; to some extent, however, this seemed to make matters worse. It became clear that the requirement for lowest-bid competitive tendering, contracts that shifted risk wholesale and a contract administration that was inflexible and adversarial might be causing some of the problems. The Latham Report was set within this climate and presented the idea that progress and success might be delivered better by teamwork and making contracts fairer (Latham, 1994). The Egan Report locked this strategy into procurement practice and created an agenda for change based on measuring performance and targets to be achieved (Strategic Forum, 1998). The Achieving Excellence in Construction initiative was launched in 1999 to promote the delivery of the best value for money by balancing quality and whole-life cost in order to meet user requirements.

Construction was taken as the first Kelly market for attention, and this has initiated a Smarter Construction Division at the OGC and the rollout of ideas through all departments (OGC, 2004). This is evident in new approaches being developed in all departments that relate to the higher-level strategy presented in the Kelly Report. The OGC set a target in 2003 that 70% of construction projects would be delivered on time, within budget, to exceed stakeholder expectations and with zero defects. For this purpose, it created procurement guides and a successful delivery toolkit. It reported on the success of this in a report, *Building on Success*, in 2003 (OGC, 2003b). In 2005, a Construction and Facilities Management Efficiency division was created in the OGC in order to support initiatives in the public sector, particularly in social housing, healthcare and education.

Government departments, including the three we have chosen, are overshadowed by this strategic political agenda for service improvement at lower cost. They are being driven to provide better value plus an encouragement for at least a better relationship with the private sector, and this dominates much of their thinking.

Defence Estates

The MOD provides political and managerial control over all UK military operations and resources for three armed forces spending some £30bn per year. Historically, the armed forces are linked to state and government power because of their role in territorial acquisition and defence against invasion. Successful military campaigns provide both identity for a nation and political and economic power in the world. This gives the military establishment an elevated position within government because defence is a political agenda with the electorate. The cost of the defence budget is high, consuming 6% of the gross domestic product (GDP); thus, this is a target for efficiency. The MOD has a very large staff base, employing about 82,000 civilian and 190,000 military personnel. The MOD maintains a hierarchical structure, whereby ministers oversee 11 delivery areas called top-level budget (TLB) holders, namely: Central, Chief Joint Operations (CJO), Land Command, Strike Command, Northern Ireland, First Sea Lord, Second Sea Lord, Adjutant General, Personnel and Training Command (PTC), Defence Procurement Agency (DPA) and Defence Logistics Organisation (DLO). These TLBs fall into three divisions: operations and front line, personnel and support (Figure 8.1).

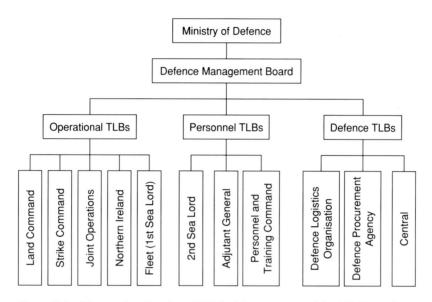

Figure 8.1   The top-level budget (TLB) holder structure of the Ministry of Defence (MOD).

Essentially, the main function of defence in the UK is to win battles in order to defend the UK and its interests. However, there is a new role associated with strengthening international peace and stability, which requires different approaches and equipment. Whereas the threat of the Soviet Union was the major issue from 1945 to 1989, where large battlefield confrontations were envisaged, the new challenges are smaller and more integrated into political problems between and within other countries. The notion of fighting men and, now, women has also changed, from a large conscripted force with limited skills to a smaller professional fighting force possessing a variety of well-trained skills. These changes in military strategy are substantive and to some extent are in conflict with military tradition. This induces uncertainty and creates anger both politically and organisationally. Practically, it involves changing the nature of weapon systems, the balance of the services, the accommodation and the service personnel. Strategic defence reviews in 1998 and 2002 set the policy for many of these changes.

The MOD is one of the largest construction clients in the UK, with an estate valued at over £14bn and an annual expenditure on landed property of over £1bn. Defence Estates (DE) is one of the MOD's units,

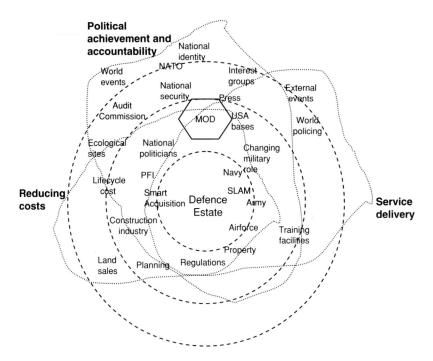

MOD, Ministry of Defence; PFI, private finance initiative; SLAM, Single Living
Accommodation Modernisation.

Figure 8.2   Map of Defence Estate's (DE) environment.

with a responsibility for looking after land and property. A map of DE's
environment is shown in Figure 8.2. DE was launched in 1994 and
became a TLB unit in April 2005. Its head office is in Sutton Coldfield,
Birmingham, and it has offices throughout the UK. DE, like the MOD,
is staffed by both military and civilian personnel, the majority of the
latter consisting of professionals and technical experts. The unit is
headed by a vice admiral.

The estate consists of about 240,000 ha of land on over 4,000 sites.
DE divides the estate into 'built' and 'rural', where built includes barracks, naval bases, air bases, hangars, accommodation and housing.
The 160,000-ha rural estate is used mainly for training. DE also manages
estates in other countries where there is a strategic military presence or
training bases. The 1998 Strategic Defence Review set demanding targets
for the disposal of land and property, which defence estates coordinated
with the expectation of £500m receipts for the period 2003–04 to
2005–06. This contraction of the estate involves the concentration of

activities into core sites. The environmental significance of an estate this size is acknowledged by the MOD, which requires DE to have a sustainability policy both for the effective management of its rural ecology and for the preservation of heritage buildings.

DE procures construction projects for the MOD, and its mission is to 'deliver estate solutions to defence needs', achieving better value for money out of the estate. DE is the MOD's projects-delivery mechanism, which interfaces with the construction world. Its response to the greater government agenda of improvement was to develop and use Smart Acquisition through public–private partnerships, including private finance initiatives (PFI) where appropriate. Smart Acquisition (see www.ams.mod.uk) is a long-term MOD initiative to improve the way it procures defence capability. The assets are no longer replaced simply by updating what was there before but involve assessing how an asset's capability will integrate with other capabilities in order to achieve optimum effect for the armed forces. This also involves adopting a whole-life approach to acquisition rather than concentrating on the initial cost of the resource. DE operationally is structured into two UK divisions, using five regionally based prime contracts – Scotland, the south-east, the south-west, central (north, Midlands and Wales) and the east – and generally procures construction work through these contracts. Large or complex projects may be procured via PFI or through stand-alone prime contracts. This is designed to align with the general government policy on value for money and better engagement with the private sector.

The politics of DE is set within the changing position of the MOD. Defence (or war) has had a major importance to governments and the nation. It is part of national identity in relation to other countries. It is also an important provider of a sense of security within the nation, and this is a major political issue at elections. The change in the purpose of the military and the change in approach challenge the political agenda as well as the services. Thus, there is significant politically derived change, and this is demonstrated in the MOD's new requirement for its estate. There is an embedded conflict between the military service and the political domain in which DE operates.

Highways Agency

The DfT oversees the delivery of a reliable, safe and secure transport system that responds efficiently to the needs of individuals and businesses while also safeguarding the environment. In 2005, the ministry had six groups: a driver vehicle and operator group; an aviation, logistics and maritime group; a delivery and security group; a roads, regional and local transport group; a rail group; and the Highways Agency. In 2004, the government published a White Paper, The Future of Transport (DfT, 2004), which discussed the role of transport, whether road, rail, bus, foot, bicycling, sea or air, in a growing economy with an increasing demand for travel set against environmental concerns. The strategy recognised the need for sustained investment, for improvements in transport management, and to plan ahead, while the quality of environmental issues were addressed by design and technology developments. The DfT took on the greater government agenda of efficiency and value for money through developing better project-appraisal techniques. The New Approach to Appraisal (NATA) is a process that explores the potential for different solutions to transport problems (see www.webtag.org.uk). NATA takes into account a wide range of factors reflecting the government's five objectives for transport: environment, safety, economy, accessibility and integration. Results are summarised in an appraisal summary table (AST), which is presented to ministers in order to inform their decisions. In addition to the monetary impacts, presented in a benefit/cost ratio (BCR), the AST also takes into account impacts that are difficult to present in monetary terms. The combination of both allows an assessment to be made of the proposal's value for money (DfT, 2004).

The Highways Agency was established in 1994 and is responsible for motorways and trunk roads in England. Local-road networks are the responsibility of local authorities. A map of the environment of the Highways Agency is shown in Figure 8.3. The Scottish Executive is responsible for motorways and trunk roads in Scotland, and Traffic Wales is responsible for motorways and trunk roads in Wales. The agenda of the Highways Agency was set by the White Paper (DfT, 2004) and involves developing and maintaining a road network enhanced by:

❏ new capacity where it is needed, assuming that any environmental and social costs are justified;
❏ locking in the benefits of new capacity through various measures, including some tolling and carpool lanes where appropriate;
❏ leading the government debate on road pricing and its capacity to lead to better choices for motorists;

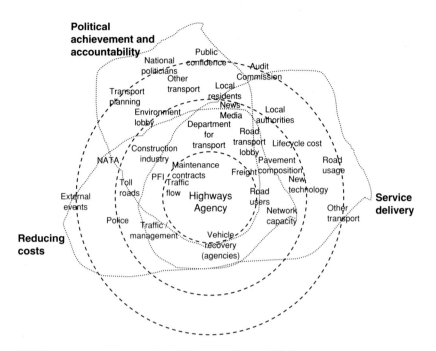

NATA, new approach to appraisal; PFI, private finance initiative.

Figure 8.3   Map of Highways Agency's environment.

❑ creating better management by exploiting the potential of new technology in order to avoid problems and deal with them rapidly if they occur;
❑ using new technology to keep people informed, both before and during their journey.

DfT (2004)

The strategic road network in England is valued at over £65bn and comprises some 4,818 miles (7,754 km) of trunk roads and motorways. In 2004, over 170 billion kilometres of vehicle journeys were undertaken on major roads, and car travel accounted for four-fifths of the total distance travelled by the UK public. National roads thus provide a vital service to commerce and industry and have a major impact on the lives of individuals and communities. As far back as 1999, road freight carried nearly 90% of total freight tonne-kilometres moved, excluding movement by water and pipeline (DETR, 1999). Through the government's 10-year plan for transport, funding of up to £22bn has been earmarked for the Highways Agency in order to improve services to customers over the decade. Over £10bn of this funding will cater for the widening of 5% of the network, which will also concern 360 miles (13%) of motorways; 80 major junction improvements, each worth over £5m; and 30 bypasses.

The politics of highways is both economic and personal. Businesses need roads to move their goods and people, such that any delays and their cost implications directly affect the operational viability of companies. There is a very strong road-transport lobby, which can get media coverage. Individuals using roads get frustrated by delays due to lack of capacity or blockages caused by poor maintenance or accidents, and this can be exploited in a political way. There are also political groups against roads because of the role of roads in pollution and environmental loss; such groups often support the development of public transport. The politics also involves road-building programmes, as these have been a traditional way for governments to use construction to invest in the economy.

Her Majesty's Prison Service

The Home Office is the government department responsible for prisons and the police force in England and Wales, national security, the justice system and immigration. It strives to provide a safe, just and tolerant society through five aims:

1 People are and feel more secure in their homes and daily lives.
2 More offenders are caught and punished and stop offending, and victims are better supported.
3 Fewer people's lives are ruined by drugs and alcohol.
4 Migration is managed to benefit the UK, while preventing abuse of the immigration laws and of the asylum system.
5 Citizens, communities and the voluntary sector are engaged more fully in tackling social problems, and there is more equality of opportunity and respect for people of all races and religions.

Home Office (2005)

In 2005, the Home Office had five operational/delivery groups: crime reduction and community safety; national offender and management services; immigration and nationality; communities; and the Office for Criminal Justice Reform.

HMPS is set within the National Offender Management Service (NOMS) group. A map of the HMPS environment is shown in Figure 8.4.

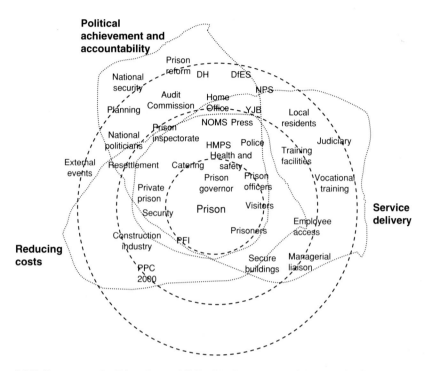

DfES, Department for Education and Skills; DH, Department of Health; NOMS, National Offender Management Service; NPS, National Probation Service; PFI, private finance initiative; PPC, Project Partnering Contract; YJB, Youth Justice Board.

Figure 8.4   Map of Her Majesty's Prison Service's (HMPS) environment.

A devolved administration works within Scotland. Prison sentences are used to punish offenders for serious offences. Incarceration both punishes offenders and protects the public, although prisons also provide reform and care of offenders. The UK prison population was about 77,000 in 2005 (Home Office, 2005). There were 139 prisons in England and Wales in 2005; most were operated publicly, but some were operated privately. The Prison Service is large, with a workforce of nearly 35,000 staff and costs over £2bn to run every year, representing about 16% of the Criminal Justice System (CJS) budget. There are also about 1,500 probation offices supported by resources such as warehousing and training facilities for probation officers.

HMPS aims to look after prisoners with humanity and to help them lead law-abiding and useful lives, both while in custody and subsequently after release. Their vision is to provide excellent prison services, hoping to achieve this through the following objectives:

❏ holding prisoners securely;
❏ reducing the risk of prisoners re-offending;
❏ providing safe and orderly establishments in which prisoners are treated humanely, decently and lawfully;
❏ providing an effective and efficient custody and escort service to the criminal courts.

(www.hmprisonservice.gov.uk)

The prison estate is diverse, with different levels of security in order to accommodate various types of prisoner. There are four main categories of prisons (Table 8.2).

The politics behind prisons relate more to the capture and prosecution of offenders than to the delivery of correctional services. There are periodic concerns that too few offenders are being incarcerated and that prison conditions are not harsh enough to deter offenders. However, the politics within the service itself are high because of changing staff regimes and, in particular, the introduction of private prisons. Traditionally, prison officers were well organised and resistant to changing their work practices to make them more efficient, which has caused some industrial unrest. The introduction of private prisons was seen as a threat by prison service managers and by prison

Table 8.2   Categories of prisons and prisoners.

Category	Descriptor
A	This is for prisoners whose escape is highly dangerous to the public, police or the state security. The aim is to make the escape of category A prisoners impossible.
B	Highest conditions of security are not as necessary for category B prisoners, but escape must be made very difficult.
C	Category C prisoners cannot be trusted in the open but do not have the resources and will to make an escape attempt.
D	Category D prisoners can be trusted reasonably in open conditions.

In addition, there is a further categorisation of prisons, with young offenders' institutes, juveniles' centres, remand centres, resettlements, holding centres, local prisons, immigration removal centres, closed prisons, semi-open prisons, prisons for males and prisons for females.

Source: http://pso.hmprisonservice.gov.uk/PSO_0900_categorisation_and_allocation.doc

officers, who saw their role being eroded further. The poor morale of staff, increasing numbers of prisoners and overcrowding within prisons at certain times have been seen as problematic by the Prison Inspectorate, and this has caused political tension.

8.3   THE MANAGERIAL DOMAIN

Decisions about how each of these departments operates are made in the managerial domain. This civil service has a high degree of rationality, both in the forms of information it produces and in how it produces this information. It uses bureaucracy and formal methods to make decisions and demonstrate accountability to its political masters. These managers need to manage the boundary between the professional/operational domain and the political domain as well as the divisions within the professional/operation domain.

The government has sought to manage departments through setting policy and then establishing targets that can be measured. Managers then establish how the targets can be met within the resources with which they are provided. The agenda has very much been to reduce the cost of operating services and to deliver a higher quality of service. The public-relations aspects of this are particularly important, although this may not be stated explicitly. The difficulties involve motivating the professional/operational staff through these changes, which involve the staff in extra work. At the same time, changes in procurement practice, including better relations with the private sector, may be resisted by some staff, who feel that these political ideas should not be introduced. The use of rational planning of these changes and the use of rational accounting practices to monitor the changes is part of the method of operation of the civil service; however, these can be manipulated to either protect current practices or to demonstrate that changed practices introduced by management have been effective.

Defence Estates

DE's head office is in Sutton Coldfield, Birmingham, and it has business units in the West Midlands, Cambridgeshire, Fife and Hampshire. The management structure is shown in Figure 8.5.

The inclusion of both military and civilian personnel allows the boundary between the political and operational domains to be

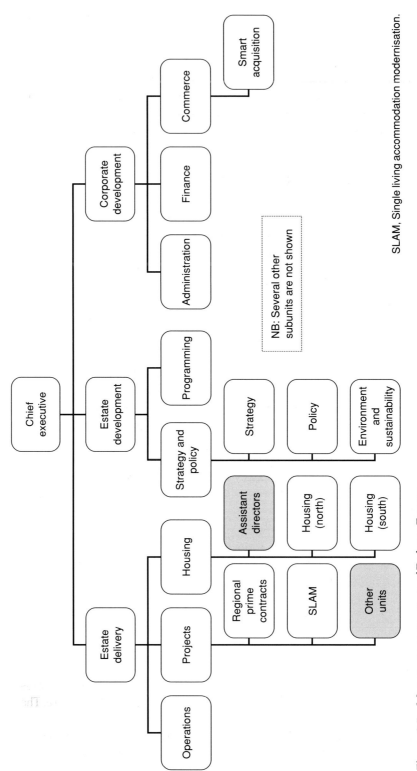

Figure 8.5 Management structure of Defence Estates.

SLAM, Single living accommodation modernisation.

managed effectively. This structure explicitly acknowledges the move to PFI and prime contracting and the use of smart acquisition in the procurement of equipment.

Since 1998, DE management has been seeking the following in order to address its political domain:

❏ To progress a large PFI programme and to roll out the MOD's prime contracting initiative as the preferred construction-procurement strategy, both for stand-alone projects and on a regional basis for core works and capital maintenance.
❏ To drive forward a modernisation programme in order to upgrade services' single living accommodation on behalf of the MOD.
❏ To achieve significant land and property receipt targets set by annual spending reviews.
❏ To put in place a new structure within the MOD in order to manage the estate so as to meet military needs more effectively and efficiently.

Highways Agency

The Highways Agency head office is in London, with regional offices in Leeds, Manchester, Birmingham, Bedford, Hemel Hempstead, Bristol, Dorking and Exeter. The Highways Agency is managed by a board of directors, which comprises the chief executive, three non-executive directors and nine executive directors. The executive directors are each responsible for one business area within the Highways Agency, which are:

❏ Procurement Directorate
❏ Network Strategy
❏ Traffic Operations
❏ Major Projects
❏ Safety, Standards and Research
❏ Finance Services
❏ Human Resource Services
❏ Corporate Directorate
❏ Information Directorate

The management structure demonstrates the agency's concerns for frontline delivery in traffic operations, which are its public face and its political connections.

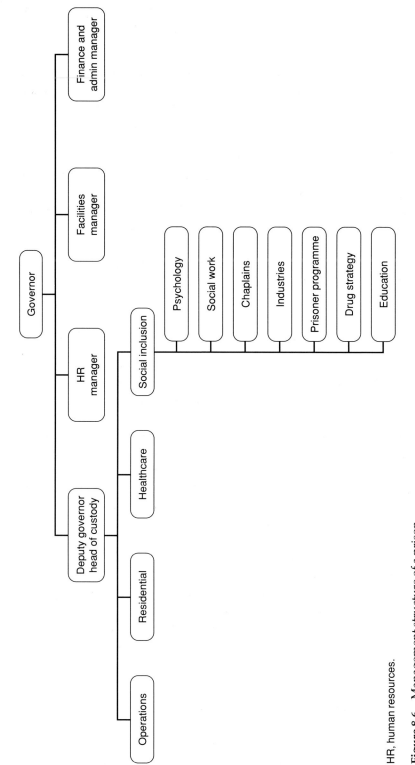

HR, human resources.

Figure 8.6 Management structure of a prison.

The agency's procurement strategy, 'Delivering Best Value Solutions and Services', was published in November 2001. Since then, the agency has changed its role to one of network operator. The Highways Agency set the target of improving efficiency in roads procurement through adding value to service delivery of £66m in 2005–06, £132m in 2006–07 and £200m in 2007–08. In addition, the agency had a change agent role to work with local authorities, with a target of achieving efficiency improvements in the procurement of services on the local road network of £60m in 2006–07 and £190m in 2007–08. The agency aimed to deliver this increasing programme within its current administration of resources by improving its business processes and reducing resources in support roles, while maintaining or, where appropriate, increasing resources in frontline delivery roles.

Her Majesty's Prison Service

The prison service in England and Wales has two management structures. The first is concerned with the strategic delivery of the service, which involves a directorate with five divisions: operations, finance, high-security prisons, personnel and prison health. The second structure involves the management of individual prisons and is headed by the prison governor; an example for a prison is shown in Figure 8.6.

The service is monitored by HM Inspectorate of Prisons, which is operated by the Home Office but outside the prison service. It provides independent scrutiny of the conditions for and treatment of prisoners and other detainees, promoting the concept of 'healthy prisons', in which staff work effectively to support prisoners and detainees in order to reduce re-offending. The inspections can be announced or unannounced and can lead to serious defects in operation or accommodation being exposed. Prisons are rated on a performance scale of 1–4. Level 4 is awarded to excellent establishments that are delivering exceptionally high performance. Level 1 indicates a poor performer.

There is a tension between the civil servants, the prison governor and staff, and the Home Office inspectorate, which moves many managerial issues into the political domain.

8.4   THE OPERATIONAL DOMAIN

The three departments are very different in what they operate and, ultimately, in the way they connect to buildings. Their differences are thus highlighted in the following sections focusing on each department's construction/procurement units as regards the declared operations of its services.

Defence Estates

Defence Estates manages houses and over 45,000 buildings for the MOD, including barracks, naval bases, depots and aircraft hangars. DE also looks after about 240,000 ha of land, located in different parts of the UK, used for 21 major and 39 minor armed-forces training areas. Service personnel tend to stay at a certain posting for two or three years, after which they are moved to other jobs. Since the personnel are itinerant, DE seeks to ensure continuity for them, aiming at providing facilities that directly address their needs. Accommodation is a critical issue with the armed forces, given that poor accommodation is among the top five reasons why people leave the services. Therefore, DE continually appraises the accommodation it provides because it wants to attract and retain professional service people.

DE has opted for larger and longer contracts, working together with industry to deliver value for money to its clients. The principal concept spearheading its construction procurement is prime contracting, which is now DE's default procurement route. Prime contracting is innovative and concerns Smart Acquisition to non-equipment procurement. Smart Acquisition is a long-term MOD initiative to improve the way it acquires defence capability. Military equipment, services, estates and business information systems are no longer replaced on a like-for-like basis but are reviewed against wider resources in order to achieve optimum effect, including adopting a through-life approach to acquisition rather than concentrating on the initial procurement cost. The MOD is operating five regional prime contracts (RPCs) to secure its core works and maintenance services across the UK. The first RPC, covering Scotland, was awarded in March 2003. The next two RPCs were awarded in March 2004 and cover the south-west and south-east of England. The fourth and fifth RPCs covering the east and central (north of England, the West Midlands and Wales) were awarded in November 2005. Supplementing the RPCs are stand-alone prime contracts, of which nine have been awarded to date. These contracts are let on a whole-life-cost basis where long-term environmental concerns are considered, especially as they have a bearing on training. Thirty per cent of the MOD's training land is in national parks, and much of its rural land has environmental and landscape significance.

The MOD thus follows thoroughly the policies of government on sustainable development, taking into consideration social, economic and environmental sustainability in its work and future plans.

A different type of contract is being developed for overseas, which is similar to the regional prime contracts. Also, DE sometimes (albeit very infrequently) uses stand-alone functional prime contracts and the PFI methodology on some of its projects. Thus, there are several opportunities to gain construction project works with DE, but the greatest of these at the moment is through the RPCs.

Respect for people and other 'soft issues', as advocated by Latham (1994) and Egan (1998), predominate the MOD's contracts, attracting a weighting of up to 80%. These yardsticks improve the value that the MOD obtains from the construction and maintenance of the defence estate. In terms of briefing, when DE gets an original list of requirements from one its TLB holders, it works out a quick estimate initially to establish how much it would cost to execute. DE then goes through the feasibility stage and comes up with a revised figure. Proceeding with the evaluation, DE moves into subsequent iterations, until a contractor is in place when a maximum target cost price is negotiated.

Highways Agency

The Highways Agency manages the national road network as a public asset as well as the vehicular traffic on it. Road construction can be new-build or restoration, rehabilitation and resurfacing (3R) (Herbsman and Glagola, 1998). A road has a design life, and part of the role of maintenance is to monitor the conditions of roads to determine whether and when interventions are needed, i.e. to repair, resurface or reconstruct a highway. This decision is based on the design life and asset condition survey of the road. The Highways Agency aims for a whole-lifecycle cost that minimises long-term maintenance costs and disruptions to users while being affordable; the latter is dependent on national and regional budgets.

In terms of delivering major projects (i.e. those exceeding £5m each), the Highways Agency's preferred philosophy for procurement is early contractual involvement (ECI), through which the contractor is engaged before the design phase. This philosophy embraces performance and supply-chain management, whereby the client and suppliers work together in communities in order to address the issues that affect everybody. A key concern is how the consortium addresses the environment. It is believed that by involving the various team members early, the potential for achieving value for money is much higher.

For medium-sized projects (i.e. those in the range £1–5m), general framework agreements are used, ranging from three to seven years in duration. Maintenance works costing less than £1m are delivered under the frameworks as task schemes. These task schemes are variants to the framework agreements, but all are based on the NEC Engineering Construction Contract standard agreement. Major schemes are tendered and bid for on an individual basis. A consortium can thus bid for a big project each time a project is advertised by the Highways Agency. For smaller projects, an organisation or consortium would have to bid to become a partner with the Highways Agency in order to tender and bid.

Delivery of best value is a major criterion for awarding project works, and this subsumes satisfaction with the product, service, defects, safety, etc. An integrated delivery team will include a designer, a contractor and any specialists or subcontractors that may be needed. Projects are executed on a cost-plus type of contract (i.e. actual cost plus overheads and profit). Teams are procured through a management-agent-contract(or) in order to manage their roads.

The Highways Agency's asset size has remained stable in the short term because, as it builds, some of its roads are de-trunked to local authorities. However, the agency has grown and changed in other ways, with the incorporation of live traffic management in its functions. This role, which used to be within the remit of the police force, has now been transferred to the Highways Agency. The size of the Highways Agency in terms of personnel has almost doubled due to its traffic-management function. As an organisation, the Highways Agency will continue to learn and change as it operates live traffic management.

Her Majesty's Prison Service

Prisons are emotional places for both the prisoners and the prison officers. Emotions are heightened when incidences occur (Crawley, 2004). In the period 1990–2001, 26 cases of prisoner-on-prisoner homicides were recorded. Victims were more likely to be housed in shared cells (Home Office, 2004). Security in prisons is prime, in order to protect inmates from self-harm and from each other. Self-harm is a key area of concern within prisons; the extreme form of self-harm is suicide, and inmates are six times more likely than people outside prison to harm themselves. Teenage inmates are at increased risk of harming themselves.

Apart from incarceration, the objectives of HMPS are to stop people from re-offending and to rehabilitate offenders with drugs problems. HMPS is thus keen to improve the education, training and healthcare of those incarcerated. There is a stated belief in the prison service that being sentenced to a prison term is the punishment itself. Life in prison, and its associated conditions, should not punish the inmates further – hence the provision of rehabilitation and other programmes designed to improve the quality of prisoners' lives during and after incarceration. As such, the service wants to provide decent conditions and environments that support prisoners before they return to the community, with the aim of making the prisoners responsible and law-abiding citizens (Neil, 1995).

The number of people in custody, including remand population, untried prisoners and convicted but unsentenced prisoners, is increasing. The population of people in custody at 31 August 2005 was 77,388 compared with a maximum capacity of 77,628 (NOMS, 2005). The prison service is having to cope with the fact that capacity is almost saturated, and finding a space in the right place and at the right time cannot be guaranteed. This operational problem becomes a political issue when events such as suicides and riots occur. Some people have been kept in police-cell custody because the local prison was full to capacity. It has been estimated that the prison population could rise to a top-end figure of 90,780 by 2011, which exceeds the current maximum capacity by far (Home Office, 2005).

In terms of buildings, prisons have holding cells and associated facilities such as farms, workshops, educational buildings, sports centres, health centres, visits' buildings, gatehouses and administrative buildings. In some low-security prisons, the sports facilities are located outside the main premises. A prison cell has sleeping, storage, washing and study facilities, and a toilet. There is a minimum size requirement for a prison cell, for which there is a design guide, but central to that is security to prevent prisoners from escaping. In addition, prisoners must be kept safe from staff, fellow inmates and visitors.

Prisons are operated in a mixed regime of public and private facilities. Private prisons have been procured under a PFI arrangement, and this approach will continue. However, publicly financed prisons will still be the main form of delivery. In the past, the public service Prison Service used a design-and-build form of procurement for its schemes. At that time, independent design consultants produced outline designs, which were novated to the constructors. In recent projects, the constructors were engaged to develop the complete designs from the brief within the wider changes of prison service procurement practice (House of Commons, 2003). HMPS contracts are now based on Project Partnering Contract (PPC) 2000. The client, contractor and consultants all sign the same contract, removing some of the interfaces and conflicts in the traditional form of contracting. There is thus a diminishing need for independent design consultants in this subsector. HMPS's brief will contain output specifications, and its property services unit will develop options to meet these in conjunction with consultants or constructors. An option appraisal is then carried out on the basis of affordability.

Since 2004, HMPS's procurement regime utilises framework agreements. NOMS has three alliances in place, with:

❑ consultants, for professional expertise (design, cost, project management, building surveying, etc.)
❑ constructors for new-build facilities
❑ constructors for refurbishment schemes.

Partners were selected through competition and allocated work, as and when this arose. Performance was monitored and benchmarked, and best practices were captured and fed through the system for overall gains. With these partnerships in place, a large amount of bidding time and associated expense was reduced. The alliances were set to each last for about four years under European Union (EU) regulations, following which another set of alliances must be established.

8.5   EXPERIENCE OF BUILDING: FROM UNKNOWNS AND CONTRADICTIONS TO MEANS AND ENDS

With each of the departments, there are uncertainties and inherent contradictions in their activities. This applies in particular to gaps between the political, managerial and operational levels and to changing political pressures. We can flag only some of these conflicts here and demonstrate how the departments are coping with them. As the industry engages with government departments, it will need to manage these conflicts; expecting them up front helps in the preparation. Means-and-ends diagrams are provided for the three departments in Figures 8.7–8.9.

Defence Estates

The relationship between DE and the military is problematic because of changes in the service and in the specific needs of accommodation. The military is used to being in command and having absolute authority. Base commanders see their role as maintaining the service against a reduction in capacity. They are also career-oriented, which requires them to make achievements in order to be promoted. Building provides the opportunity for these activities, and they can try to influence developments both formally and informally.

There is unlikely to be money available to carry out all of the desired works, and so priorities have to be made. Some built facilities are upgraded ahead of others. Therefore, solving a problem in one location creates a demand in another location, which puts a strain on the system. This is an inherent conflict that raises emotions of frustrated expectations, which are difficult to manage.

There could be unknowns in a brief due to changing military practice. Although these unknowns are prevalent, they are handled through rational judgement for accountability. For instance, DE's risk process is rationalised by using approaches such as Monte Carlo simulation. This risk-assessment tool treats the unknowns seemingly rationally. A risk-averse attitude, like DE's, uses rationality; however, rationality is often bounded, and the end may meet the accountability criteria but not the operational criteria.

As a means of achieving more effective ends, DE has turned to long-term contracts (partnerships) (see Figure 8.7). These are output-based and enhance the upgrading of facilities, e.g. from grade D to grade B. As such, for example, the contractor can decide to re-roof a building now rather than patch it for the next seven years.

Under prime contracting, work is outsourced to a private prime contractor; thus, some of the risks and accountability sit with these contractors. If something goes wrong with the functioning of a facility, then it is the responsibility of the contractor to fix it within an agreed timescale, ensuring that services are not disrupted for long. On the maintenance side, DE's long-term relationships are used as a means to guarantee the operations of facilities and the delivery of accountability.

Thus, the industry has to manage:

❏ conflict resulting from the change in services organisation
❏ career aspirations and role values of senior military personnel

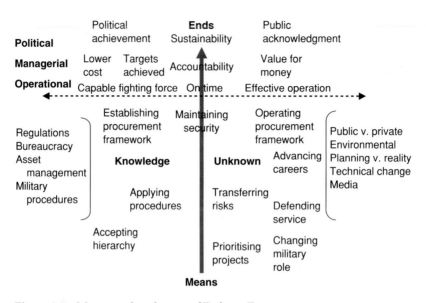

Figure 8.7   Means-and-ends map of Defence Estates.

❑ conflict with resource limitations
❑ changes in military practice
❑ military need for rationality.

Highways Agency

The Highways Agency has a conflict of purpose set within its direct exposure to the public and its roads. There is a long-term problem with regard to the environmental impact of roads and whether to develop them at all or whether to constrain their use. There is a short-term problem of keeping traffic moving. This conflicts with the long term but is also problematic in the short term, as maintenance requires roads to be closed or their use restricted. This results in public complaints.

The Highways Agency tries to rationalise this into procedures and uses risk management as a part of its project management. The agency in particular evolves a risk register for every scheme, and risk-management workshops are held to determine courses of action. Risks on road projects could pertain to the ground conditions, protests against a certain route, etc. Risks also inform budgeting. If risks are anticipated, then their cost impact is reflected in the budget, which the Highways Agency hopes protects it from complaints.

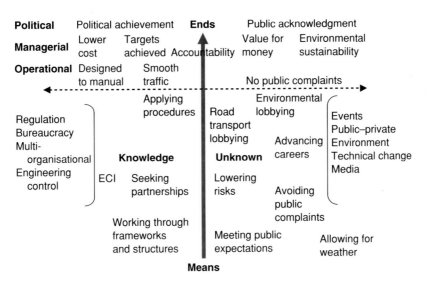

ECI, early contractual involvement.

Figure 8.8 Means-and-ends map of the Highways Agency.

At the project team level, a cultural assessment framework guides the relationship between the members. The framework encourages teams to address issues that may arise to undermine the progress of the team. In terms of the products and services, structures are in place to determine what success is and how successful they have been (i.e. leaning towards bureaucracy). In terms of performance management, the framework agreement removes subjectivity while allowing flexibility (Figure 8.8).

Thus, the industry has to manage:

❏ conflict of purpose, i.e. traffic flows or sustainability
❏ conflict of action, i.e. frequent short-term disruption or one-off long-term disruption
❏ change in public expectations
❏ the Highways Agency's need for rationality.

Her Majesty's Prison Service

The prison service (Figure 8.9) has a conflict of operations because of limited capacity and changing requirements for the delivery of

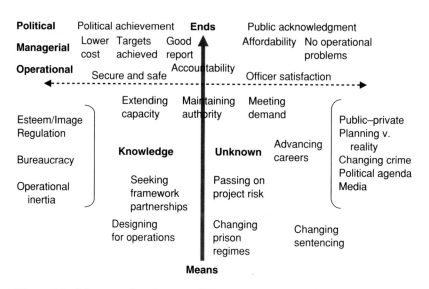

Figure 8.9   Means-and-ends map of Her Majesty's Prison Service.

incarceration. The service is required to accept the prisoners that the judicial service gives it. The variability in this, in both day-to-day operations and long-term capacity, are problematic. These problems could be passed on to the private sector in order to manage the risk. However, this solution is limited by the costs and the potential of failure being passed back to the prison service.

The prison population is unwilling and volatile, with the risks of disorder, violence and self-harm being ever present. There is an increasing rate of suicides in prisons. HMPS liaises with its designers to come up with solutions that do not facilitate suicides or hazards. There is a hope that this risk can be managed by better building design in order to integrate it with prison management. The industry has to deal with a changing service while delivering new and refurbished facilities.

Thus, the industry has to manage:

❏ conflict between the Home Office and the prison service
❏ changing security perceptions of the service
❏ conflict with resource limitations
❏ changes in prison practice
❏ service need for rationality.

8.6   KEY ISSUES

A government requires better service delivery at lower cost but with substantial accountability. In any government construction project, fours sets of changes occur:

❏ the construction project
❏ the organisational improvement construction produces
❏ the procurement mechanism set by government's overall agenda
❏ the departmental role and processes set by the government's overall agenda.

This degree of change and the contradictions between the political, managerial, and operational domains is a challenge to the people and the system. The response to construction may not be understandable from a construction perspective alone.

Government departments have embraced modern forms of procurement and are keen to obtain better value from their facilities. There is an element of trust involved in this change, i.e. that construction supply chains will deliver what these departments want in these four change situations. The relational types of agreement do allow a better basis for understanding these clients: by engaging the supply chain early, the opportunity to evolve clients' briefs together and to study clients' needs is greater. Also, by working with a government client over time, construction supply chains get to know their clients better, removing the personality and cultural barriers that hamper communication.

REFERENCES

Crawley, E.M. (2004) Emotion and performance: prison officers and the presentation of self in prisons. *Punishment and Society*, **6**(4), 411–427.

DETR (1999) *National Statistics Transport Facts*. London: Department for Environment, Transport and the Regions.

DfT (2004) *The Future of Transport*. London: Department for Transport.

Egan, J. (1998) *Rethinking Construction: A Report of the Construction Task Force to the Deputy Prime Minister, John Prescott, on the Scope for Improving the Quality and Efficiency of UK Construction*. London: Office of the Deputy Prime Minister.

Herbsman, Z.J. and Glagola, C. R. (1998) Lane rental: innovative ways to reduce road construction time. *Journal of Construction Engineering and Management*, **124**(5), 411–417.

Home Office (2004) Prisoner-on-prisoner homicide in England and Wales Findings no 250. London: Home Office.

Home Office (2005) *Updated and Revised Prison Population Projections 2005–2011.* Home Office Statistical Bulletin. London: Home Office.

House of Commons (2003) Modernising Procurement in the Prison Service. Report by the Comptroller and Auditor General. HC 562. London: The Stationery Office.

Latham, M. (1994) *Constructing the Team: Final Report of the Government/Industry Review of Procurement and Contractual Arrangements in the UK Construction Industry.* London: The Stationery Office.

Murray, M. and Langford, D. (eds) (2003) *Construction Reports 1944–98.* Oxford: Blackwell Science.

Neil, C. (1995) System kitchens: improving catering quality and value for money in HM Prison Service. *Facilities,* **13**(12), 21–25.

NOMS (2005) Population in Custody: Monthly Tables – August 2005 – England and Wales. London: National Offender Management Services.

OGC (2003a) Increasing Competition and Improving Long-term Capacity Planning in the Government Marketplace: OGC Report to the Chancellor of the Exchequer (the Kelly Report). London: Office of Government Commerce.

OGC (2003b) Building on Success: The Future Strategy for Achieving Excellence in Construction. London: Office of Government Commerce.

OGC (2004) Guidelines on Factors that can be Considered when Trying to Reduce the Risks of Over-dependency on a Supplier. London: Office of Government Commerce.

Office for National Statistics (2005) *UK 2005: The Official Yearbook of the United Kingdom of Great Britain and Northern Ireland.* London: Office for National Statistics. www.statistics.gov.uk/yearbook.

Strategic Forum (1998) *Accelerating Change: A Report by the Strategic Forum for Construction.* London: Rethinking Construction.

RESOURCES

Government: www.direct.gov.uk
Treasury: www.hm-treasury.gov.uk
Office of Government Commerce: www.ogc.gov.uk
National Audit Office: www.nao.gov.uk
Ministry of Defence: www.mod.uk
Defence Estates: www.defence-estates.mod.uk
Ministry of Defence Acquisition Management System: www.ams.mod.uk
Her Majesty's Prison Service: www.hmprisonservice.gov.uk
Prison Reform Trust: www.prisonreformtrust.org.uk
National Offender Management Service: www.noms.homeoffice.gov.uk
Department for Transport: www.dft.gov.uk
Highways Agency: www.highways.gov.uk
Transport 2000 lobby group: www.transport2000.org.uk

9 Airports as Clients

9.1 INTRODUCTION

Travel has always commanded the attention of people, but the rise in air travel since the 1960s has been dramatic, with the rapid and distant moving of goods and people at a low cost. This has changed the world. Although information technology (IT) has been identified as the precursor to globalisation (Castells, 2001), air travel has given it physical meaning. We now live in a small world where everywhere is just a day away. Air travel has moved from being a luxury for the few to being a necessity for many; it is now a mode of public transport and is part of the political discussions about economic development of countries and regions. This makes it a major market now for the construction industry but also a large potential market as air travel and airports continue to expand.

The advance that enabled this to happen was in aircraft design and construction. Since the first flights of the Wright brothers in 1902, the aeroplane has been a classic example of scientific and technological development: hypothesis, calculation, and trial and error. Whether wings, materials, engines, steering, landing gear, pressurised cabins or automated landing, each advance has been driven by a competitive desire to make it better, faster, bigger or more comfortable. Arguably, the biggest and most rapid advances occurred during the Second World War, when the drive was strongest and the risks taken could be higher than in peacetime. This caused the many aircraft manufacturers to be reduced significantly, until today when there are very few. Aircrafts are probably the most highly complex technologies used as everyday objects.

Many airports have a history that started before the Second World War as local authority or privately owned airfields. Air travel then mimicked sea travel, being luxurious and exclusive, but competed with it in terms of speed. The Second World War required extensive developments to provide landing and maintenance facilities for the airforce. As military decommissioning took place after the war, many aircraft, pilots and airports were to some extent redundant and available for civilian use. The potential of air travel, particularly over large continental masses such as the USA, was evident, and developments in scheduled air routes for business expanded rapidly. As this business traffic increased and aircraft developed, the cost of air travel reduced, and it became available to greater numbers. General economic prosperity in the 1950s fuelled a desire for leisure travel. This was exploited first in the package holiday using chartered flights, which started in the late 1950s, and then in lower-cost transatlantic flights in the 1970s. There was always a tension with these developments, as air travel was seen in strategic national terms, such that routes were controlled by governments and airlines had to apply for permission to fly. In addition, many airlines were operated as national airlines, being funded and operated by government; these airlines fought against the introduction of cheaper and less regulated air travel (Gunston *et al.*, 1992). As governments saw these national airlines losing money, they began to seek to offload them to the private sector (Graham, 2003).

Following deregulation in the 1980s, the 1990s saw the introduction of no-frills airlines, which took the entrepreneurial experience from early mass flyers and combined it with a totally re-engineered business model of airlines. Thus, the typical low-cost carrier operates by having a single passenger class and unreserved seats for quick boarding; using only one type of aircraft (thus reducing training and servicing costs); having a simple fare scheme that rewards early reservations (yield management); flying to cheaper, less congested secondary airports; concentrating on short flights and fast turnaround times to allow maximum utilisation of planes; emphasising point-to-point transit; employing direct sales of tickets and thus avoiding commissions; having employees working in multiple roles; and eliminating free in-flight refreshments. The impact of the no-frills airlines has forced prices down on all routes and put pressure on national airlines to operate more cheaply. This has further increased the use of air transport, with an increase in no-frill traffic in the UK from 8m in 1998 to

35m in 2002 (DfT, 2003). Overall, there has been a five-fold increase in air travel since the 1970s, with currently about 220m passenger journeys through UK airports (DfT, 2003).

At the same time as the increase in passenger traffic, there has been a doubling of freight traffic at UK airports from 1990 to 2003, when it amounted to approximately 2.3m tonnes (DfT, 2003). Air transport provides a fast and efficient carrier service, particularly for high-value commodities, perishable foods and spare parts for specialist engineering. Much freight is carried by passenger airlines (bellyhold), but there are many dedicated freight-only airlines and some airports that have a high volume of this freight traffic, e.g. Nottingham East Midlands.

Airports exist at three national levels: hub airports, regional airports and local airports. Traditionally London Heathrow was the central focus of air transport in both the UK and Europe, and was the third busiest airport in the world (see Table 9.1). Around 50% of air transport demand in the UK is to the London airports, and this means that they can offer many destinations and frequent flights (DfT, 2003). This encourages people from all over the UK to use Heathrow as a hub airport. However, the increase in traffic has meant that it is economical for airlines to start using regional airports. There are now about 22 hub and regional airports in the UK; the top 10 are (AOA, 2005) shown in Table 9.2. Only seven of these hub and regional airports support inter-continental flights to hub airports mainly in the USA, but all offer a large selection of European destinations. The position of the UK means

Table 9.1 Passenger traffic at the world's busiest airports.

Rank	Airport	No. of passengers (m)
1	Atlanta	83.5
2	Chicago	75
3	London Heathrow	67
4	Tokyo	62
5	Los Angeles	61
6	Dallas/Fort Worth	59.5
7	Frankfurt	51
8	Paris Charles de Gaulle	51
9	Amsterdam Schiphol	42.5
10	Denver	42

Source: AOA (2005).

Table 9.2   Passenger traffic at the UK's busiest airports, 2004.

Rank	Airport	No. of passengers (m)
1	Heathrow	67
2	Gatwick	31
3	Manchester	21.5
4	Stansted	21
5	Birmingham	9
6	Glasgow	8.5
7	Edinburgh	8
8	Luton	7.5
9	Nottingham East Midlands	4
10	Belfast International	4

Source: AOA (2005).

that it is an ideal hub for flights from Europe, the Middle East and the Far East to the USA. Regional airports deliver more convenience but cannot offer the range or frequency of international flights. With regard to the no-frills airlines, regional airports offer lower landing charges and faster turnarounds. The smaller airports, of which there are about 48, cater for charter traffic and business traffic (A complete list of airports and airfields is available at www.aircraft-charter-world.com). The term 'general aviation' involves small private operations, including flying clubs. All local airports and most regional airports support general aviation traffic and the businesses that organise them.

Most larger UK airports have been in public ownership. Although all have been forced to privatise, at least partially, as part of the Airports Act 1986, many have a local-authority presence in their ownership and will have this on their consultative committee. Privatisation has allowed airports to seek commercial funds and to expand rapidly in order to meet the needs of the developing market. The largest airports, Heathrow, Gatwick and Stansted, are owned by BAA, which was once a government corporation but is now completely privatised. A number of airport operators own or partly own a number of airports around the world (e.g. TBI, Aer Rianta, BAA).

The purpose and physical nature of airports mean that they must be monopolies operating through permissions granted by the state. Effective competition must be a considerable distance away, and these also compete for permissions. The need for the skies to be controlled for safety by an external body, now the National Air Traffic

Control Services (NATS), is viewed strategically by the Civil Aviation Authority (CAA). These organisations effectively limit the strategic and operational desires of airports. NATS was set up in 2001 as a public–private partnership between the Airline Group, a consortium of seven UK airlines, which holds 42%, NATS staff, who hold 5%, UK airport operator BAA plc, with 4%, and the UK government, which holds 49% and a golden share. Local airports may have their own air-traffic control organisation, but this needs to operate within the wider regulated and accredited system. The CAA regulates the economic operation of all airports with an annual turnover of more than £1m to cap aviation charges. It regulates Heathrow, Gatwick, Stansted and Manchester, which are 'designated airports' (CAA, 2004), more heavily. There is a move to change the CAA's role to being about arbitrating between airlines and airports around efficiency and service benchmarks (CAA, 2005).

Aviation generates about £14bn to the UK gross domestic product (GDP) and supports 675,000 jobs (AOA, 2005). It plays a significant part in UK exports (£13bn; AOA, 2005) and in the UK inbound tourist industry (estimated as worth 4.4% of GDP; DfT, 2003). Only air routes in Scotland and to the Scottish islands receive a direct subsidy, although many environmentalists complain that the lack of taxation of air fuel is a subsidy. Airports enhance the economic development of particularly local economies by encouraging businesses to locate near them. The negative issues surrounding the aviation industry concern the environment because of the industry's use of fossil fuels and the polluting effect of engines; however, it is probably the local effects of noise and the blighting of the land that generates most opposition. A fear is also generated, as airports and airlines are security targets. Aspects of local (or national) pride and the excitement of travel probably sustain airports and airlines, however, against the problems they have to face.

9.2   BUSINESS ENVIRONMENT OF AIRPORTS

Although airports have a public-service background and many are still partly owned by local authorities, they operate as private commercial organisations. They are best modelled through business systems hierarchy theory, identifying strategic, tactical and operational levels (see Section 3.3), although their strategic level has much political interaction. In addition, international and national regulations affect all

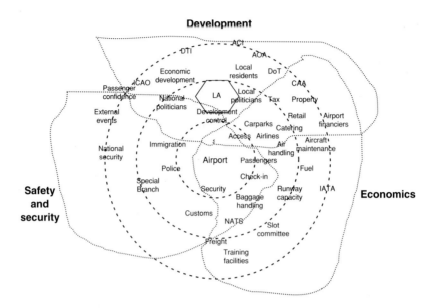

ACI, Airports Council International; AOA, Airport Operators Association; CAA, Civil Aviation Authority; DoT, Department of Transport; DTI, Department of Trade and Industry; IATA, International Air Transport Association; ICAO, International Civil Aviation Organisation; LA, local authority; NATS, National Air Traffic Services.

Figure 9.1 Map of airport business environment.

levels of airport business, some of which is political, such that there are major constraints on developments, actions and activities. Airports use their trade body, the Airport Operators Association (AOA), to lobby at this regulatory level, but there are many other lobbying groups that they can join or have to contend with. A map of the business environment of an airport is shown in Figure 9.1. This map shows areas relating to development, airport economics, and safety and security, which relate to strategic, tactical and operational concerns respectively.

Airport business

Business is a tactical issue, as planning of operations needs to take place several months in advance in order for airlines to publish their schedules and sell tickets. Airports operate and compete in two markets: the regional market (or home market), for end destination or point to point travel, and the transfer market (Graham, 2003). The former involves passengers and freight that have the region around the airport as an origin or a destination. In many regions, passengers have a choice of airport, and so there is competition between airports. Passengers are interested in:

❏ what the airlines offer at the airport, e.g. number of destinations, convenience of departure times, frequencies of departures, value of fares
❏ how easy is it to get to the airport, e.g. travel time, options of transport, quality of road or public transport
❏ accessibility of the airport arrangements, e.g. carparking regime, carparking prices, car rental, location of public transport terminal
❏ what else the airport offers, e.g. tax-free shopping, restaurants, casino.

The hub-and-spoke system in aviation means that passengers fly via a hub before transferring to a flight to their final destination. There are four intercontinental hub airports in Europe: Charles de Gaulle, Heathrow, Frankfurt and Schiphol. Traditionally, London Heathrow was the central focus of air transport in both the UK and Europe, being the third busiest airport in the world. The origins and destinations of flights can be numerous, depending on what the airport and airlines offer. Competition takes place with other hub airports, but transport arrangements and accessibility are not important; however, the convenience and ease of the connection between the incoming and outgoing flights are important. Thus, interconnected timetables and reliability of airlines are as important as the airport facilities, although in multi-terminal airports there is the problem of distance of transfer. The transfer market is a more competitive market, since people travelling from an airport often have a choice of hubs at which to transfer. However, a large home market at an airport allows higher frequencies, thus giving it a competitive edge. This increase can be exponential: as an airport attracts more passengers, it can offer more destinations and flights, and thus it attracts even more people and airlines (Bruinsma *et al.*, 1999).

Airport operations are characterised by quanta, so an additional runway creates a discrete increase in capacity. Increasing capacity is extremely expensive but, once complete, allows airports to grow with relatively low costs. A related phenomenon is the presence of economies of scope for airlines, whereby larger airlines can offer more routes with the same overheads as smaller airlines. Airlines and airports are in a symbiotic relationship; however, airports are urgently looking for opportunities to increase their numbers of customers. Thus, airports give new airlines discounted aviation rates in the hope of developing traffic and establishing routes.

As well as favourable aviation rates, airlines are concerned about operating slots, i.e. takeoff and landing times. The airport would like to smooth aircraft movements throughout the operating day, but airlines have preferred times due to passenger demand and the efficient use of aircraft stock. The allocation of slots is done under a 'grandfather rights' arrangement, with negotiations taking place between established airlines (Graham, 2003). European regulations, following International Air Transport Association (IATA) recommendations, govern the allocation of slots, allowing preferences to new entrants for any new or unused slots. An independent company has been established that coordinates the allocation of slots at larger airports. However, a market in slots has probably existed for some time, and there are moves to make this official.

The economics of airports are different depending on whether the airport is an international hub, a regional airport or a local airport. Revenue comes from either aviation (aeronautical) or commercial (non-aeronautical) sources. The aviation revenues come from the direct operation of the aircraft and include landing fees, passenger fees, aircraft parking fees and other aircraft fees that might apply. The other aircraft fee category involves items such as airbridges, push-backs, air-traffic control, slot management, fuel, ground-handling, hangers and engine testing, and these may differ at different airports or be categorised differently (Graham, 2003). The commercial revenue comes from carparking, retail concessions, catering, property rents and other business services, such as meeting accommodation.

As airports have become businesses rather than public services, the revenue from their commercial activities has increased. Thus, at the larger London airports (Heathrow, Gatwick), the aviation derived revenue is only about 40%; at regional airports such as Birmingham and Glasgow, the aviation derived revenue is about 60%; and at smaller airports it can rise to 80% (Graham, 2003). These figures represent the stress on airports to perform in different ways. It is clear that it is not possible to run airports purely from aviation revenue, and thus smaller airports are at a disadvantage.

Larger airports have a conflict of purpose, i.e. whether to invest in aviation improvement or to invest in commercial development. The dilemma has to recognise that the commercial activities require the aviation traffic. Once this is established, however, there is a temptation to concentrate on enhancing profits through commercial means. Small airports have the problem that they cannot command sufficient traffic to justify extensive retail and business facilities. Their revenues are much more tied to the airlines that operate from them. If these airlines choose to leave to use another airport, then this can induce severe financial problems. Traffic expands only with the establishment of scheduled routes, which requires the airport negotiating with airlines, which can use their power to seek major aeronautical fee reductions to use the airport. This also happens at larger airports, but the significance of the aeronautical fees in the overall economics of the airport is less in this case.

Airport development involves high capital expenditure with a long-term return. It is difficult for small airports to get the capital resources required to undertake this. In order for an airport to start to operate for commercial airlines, its facilities must reach a minimum standard for passenger expectation; however, investment in this is expensive initially, with low return, cash-flow difficulties and income variability, as business has to be built up. It is difficult for new entrants in the market, and existing airports have something of a monopoly. The overhead of having security, baggage-handling, control agencies and terminal facilities means that there is a minimum viable size. There is a status in reaching 1m passengers a year, but as regulations and thus overheads increase, even this may not be viable.

The catchment area for airport viability is crucial. For example, Birmingham International Airport has a catchment of 7m people (BHX, 1995). Local airports are selling their convenience of access, whereas international hub airports are selling their choice of destinations (Graham, 2003). This means that local airports will start to compete with each other and regional airports, depending on travel distances and flight destinations. In the current market there is a differentiation between business passengers, who require convenience and are less cost-sensitive, and leisure passengers, who are cost-sensitive but willing to suffer inconvenience of place or time. In small airports, the connection between the airline and the airport is vital. It is not surprising that a number of airports are owned by the same company that runs an airline, e.g. Coventry Airport is owned by TUI, which operates Thomson, Plymouth Airport is owned by Sutton Harbour Holdings, which operates Air South West, and Kent International

(London Manston) Airport is owned by Planestation, which operated EUjet. The economic vulnerability of small airports is exemplified by the fact that in the summer of 2005, Planestation and EUjet went into receivership, closing Kent International Airport. The ability of smaller airports to command sufficient traffic is highly dependent on competition in the region, destinations offered and airport efficiency. There is great pressure to try to undercut rivals in order to establish a market position, but this makes the business vulnerable to variability in traffic to maintain cash flow.

Airport development

The high capital cost of development and the political/regulatory environment within which this takes place make airport development a long-term strategic issue. The national and local political arenas become the ground on which airports have to seek support. Although finance is an issue, once the political environment has been moved to support the airport, then the potential economic rewards are high enough over an extended time and with low enough risk because of the monopoly character of airports to be very attractive to financiers. The risk is greatest at local airports, where the air traffic may not transpire and the political opposition is more divisive than at regional airports, which have an established airline base. The rewards of an airport in regenerating a local economy that makes it politically acceptable come about only at a certain level of traffic, which induces confidence in their development. Against this, airports are seen negatively in environmental terms and so they need to demonstrate that they operate in a responsible way. Thus, all major and regional airports have in place extensive environmental policies to demonstrate their responsiveness.

The benefits to the local economy are both direct and indirect. Directly, an airport employs a large number of people, for example Birmingham has 6,500 jobs at the airport and 1,000 offsite. Indirectly, the presence of an airport encourages companies to relocate there because of access to international suppliers, customers and markets; access to other parts of the company; and access to air freight, where security and speed are important. In addition, the region is opened up for tourist and business-convention visitors, which further enhances others' knowledge of the potential of the region. These are long-term developments locally, which are tied to longer-term economic developments internationally.

The environmental constraints surrounding airport operations that need to be managed are planning, surface access, noise, vortexes, water and air pollution, and ecology. The issues surrounding these are outlined below. As well as these issues, many airports seek to have energy- and waste-management strategies that present a positive image and save money.

All commercial airports with over 20,000 passenger-carrying movements a year are required to produce an airport master plan (DfT, 2003). This is an enabling document that can be adopted by local planning authorities into their area action plans as part of the Planning and Compulsory Purchase Act 2004. Many airports use extensive public consultation and have set up independent consultative committees together with regional economic and social analysis to set out and justify the prospects for growth of the airport and its infrastructure, including surface access. The impact of the airport on surrounding buildings and where compulsory purchase may have to take place will be made evident. Such documents are open to attack but once accepted can be used to justify future action.

As airports are transport interchanges, surface access to an airport is critical for its operational success and to avoid excessive disturbance to the surrounding communities. One issue that is being promoted is the increase in the use of public transport to access airports. Currently, even in airports with a railway station, such as Birmingham, passenger access by car is 70% and by taxi 17%, with only 10% by rail and 3% by bus (BHX, 2005). Airports monitor this and attempt to improve access by public transport; however, when passengers are using air travel to reduce travel times and reduce stress, public transport has a difficulty in meeting their needs. The development of a road system will still be a major issue with most larger airports, and funding infrastructure improvements as a way of making easy access becomes a selling point to passengers. In addition, the provision of carparking will be important, particularly as it is a significant revenue stream, and the location and management of carparks will form part of an airport's strategy.

Aircraft and the associated activities of the airport produce noise at a volume that is disturbing to surrounding communities. Some technical issues surrounding noise are given in Box 9.1. This noise comes from aircraft arriving and departing from the airport, from aircraft taxiing, from aircraft engine testing, and from the use of electrical generators. The latter involves the running of the aircrafts' auxiliary power units

---

Box 9.1   Describing noise.

Noise is described as sound pressure level (SPL) in units known as decibels (dB), which are weighted logarithmically to reflect the reaction of the human ear to the loudness of different frequencies. Maximum level (L_{max}), sound event level (SEL) and effective perceived noise level (EPNdB) are used to measure single aircraft noise events. Average aircraft noise exposure is described by the equivalent continuous noise level (Leq), which is calculated over a given time period. In the UK, average daytime noise is expressed by the 16-hour Leq (Leq_{16h}).

---

on the ground. Noise abatement is often a subject of the planning permission of the airport. One of the major issues is night flying, which is banned at some airports. Most airports place restrictions on night flying in terms of the number and the types of aircraft operating during the night period. Airlines may be given a quota or have to pay surcharges for landing at night. Night noise levels are expected to be below 87 dB(A), and day levels need to be below 92 dB(A). Birmingham International Airport surcharges airlines £500 plus a further £150 for every decibel above this level, placing the revenues in a community trust fund to support local projects.

In order to reduce the impact of noise, aircraft are expected to take off and land using noise preferential routes (NPRs). This track-keeping takes aircraft over the least populated areas, and aircraft deviating from it can be subject to surcharges. Airports map out the noise contours (66 Leq_{16h}) of these routes in order to establish the number of properties affected. In addition, the airport may fund soundproofing to affected properties. Airports are sensitive to complaints and may have an airport noise operations monitoring system (ANOMS), which uses fixed noise monitors to record noise levels combined with radar to locate offending aircraft.

Vortexes are circulating currents of air created by the aircraft as it moves through the sky. In certain weather conditions, vortexes from landing aircraft can reach ground level before they have been broken up, causing the movement and slippage of roof tiles.

The operation and maintenance of aircraft pollutes by discharging oil, glycol-based de-icing agents for aircraft and potassium acetate-based de-icing agents for aircraft stands, aprons, taxiways and runways. Other chemicals such as herbicides and cleaning agents are also used.

Oil interceptors are used for surface water runoff from operational areas. Larger airports have an inline pollution monitoring and control system.

The National Air Quality Strategy (NAQS) covers six air pollutants – benzene, carbon monoxide, nitrogen dioxide, ozone, fine particles (PM10) and sulphur dioxide – together with the hydrocarbons toluene and xylene. Larger airports monitor these themselves for protection against complaints, which can bring in local environmental health officials or the Environment Agency.

Airport sites cover several hundred hectares (Birmingham has 330 ha) and affect the environment of the surrounding rural and built-up areas. The noise, chemical pollution and planting height restrictions disturb offsite and onsite ecologies. Wildlife, particularly birds, onsite can endanger aircraft, and so these dangers have to be eliminated. The management of the grasslands bordering runways and other operational areas is important for the safety of aircraft. The effects of noise on communities can be reduced by the creation of noise bunds (high banks of grassed earth) and tree-planting barriers, and these may allow for the establishment of a new ecology. However, airports find it difficult to maintain ecology and so may compensate by creative combinations of soft and hard landscaping on the airport site, with decorative planting around the airport entrance and terminal buildings.

Safety and security

Against all the complicated strategic and tactical issues surrounding airports, airline safety and security take precedence in operational management. Air crashes and air incidents (air prox) can not only kill large numbers of people but also close airports and produce long-term negative publicity and disincentives to flying. Aerodrome safeguarding is the process whereby all measures necessary are taken to ensure the safety of aircraft, and thereby passengers and crew, while taking off, landing or flying in the vicinity of an aerodrome. The safe operation of aircraft requires accredited maintenance facilities, the handling of fuel, and the servicing of electronic communication and navigation equipment. Many of the safety systems are protected with dual provision of equipment and electrical generators. Safety systems are continuously tested and failure modes searched for. There is an engineering culture whereby all problems are investigated, documented and reported to the wider aviation community.

Of all the issues surrounding airports, however, security has been elevated in importance more than any other. The issues of terrorism, in particular the 11 September 2001 attack on the World Trade Center in New York, using a full passenger aircraft as a weapon, has changed the development and operations of airports. The ramifications of this have not been worked through fully, and they will continue to induce change for many years. Security has also become a necessary part of a reviving of public confidence in aviation for national and international tourists and businesspeople in order to maintain the aviation market and national economy.

Airport and airline security is an international political issue that is governed by international conventions (Annex 17 of the Chicago Convention 1974), International Civil Aviation Organisation (ICAO) regulations, European Union (EU) regulations, national frameworks and local operations. Apart from the scale of the 11 September 2001 attack, it was the fact that it was totally unexpected that caused the major rethink of air security. The aspects of suicide terrorism and the use of everyday items as weapons required a major change in the approach. Airports and aeroplanes are also targets for hijacking, sabotage and missile attack. Airports are either the sites of this, or the proponents pass through airports, thus making airports the most suitable places to stop these activities.

The issues of security go far beyond terrorism. Airports are like all ports, in that there are opportunities for state-proscribed and anti-social activities. The UK has a National Aviation Security Programme, as part of a wider European plan, which was reviewed following 11 September 2001 (DfT, 2002). This review looked beyond terrorist and national security issues and investigated other criminal activities and interference that might take place at airports. The security issues at an airport are detailed in Table 9.3. The control authorities (police, Special Branch, customs, immigration) have responsibility for the first two columns in Table 9.3; the third column is the responsibility of the airport. Depending on the nature of an incident, the military may also be involved. There is now a move to integrate all security issues across control authorities and across the airport functions. Thus, the use of access control with its searching and screening activities becomes part of a wider database that links with passport information, police records, Home Office records, Special Branch records, baggage records, facial recognition on closed-circuit television (CCTV), and fingerprint and other biometric data in order to provide as secure a facility as possible.

Table 9.3   Security issues at an airport.

Border controls	National security	Access control
Smuggling	Terrorism	Environmental protestors
Human trafficking	Sabotage	Unauthorised people
Organised crime	Criminal damage	Prohibited items
Illegal immigration	Theft	Inconsiderate actions

Source: DfT (2002).

9.3   BUSINESS STRUCTURE PROCESSES AND OPERATIONS

This is a highly organised world that operates through structured rational processes and is required to be planned far in advance, with deviations from these plans being experienced as problems. Many activities are set by regulation imposed from the outside, and the role of management is to ensure these happen and are coordinated across the various functions of an airport. The diagram in Figure 9.2 shows a typical management structure for a regional airport. The airport requires many other organisations, which are coordinated by the airport management, such as control authorities (police, Special Branch, customs, immigration), outsourced functions, such as air-traffic control, retail and carparking, and external agencies, such as highways, transport operators and local authorities.

Airport business requires moving the greatest number of passengers through the airport in such a way as to not delay flights, and yet giving passengers access to retail and other services that are income-generators. Although viability and operation are the main concern of most local and regional airports, a few, in particular the international hub airports, have capacity problems. Indeed, the future problem of capacity because of the demand for air travel was one of the instigators of the government White Paper 'The Future of Air Transport' (DfT, 2003). Regional airports currently may not suffer an absolute capacity problem, but many regional airports have times of the day during which the airport is at capacity. This is induced by the passenger needs and the operational requirements of airlines. For people in Europe, it is possible to arrange business meetings anywhere within Europe and to return home within a day. This induces a high demand for

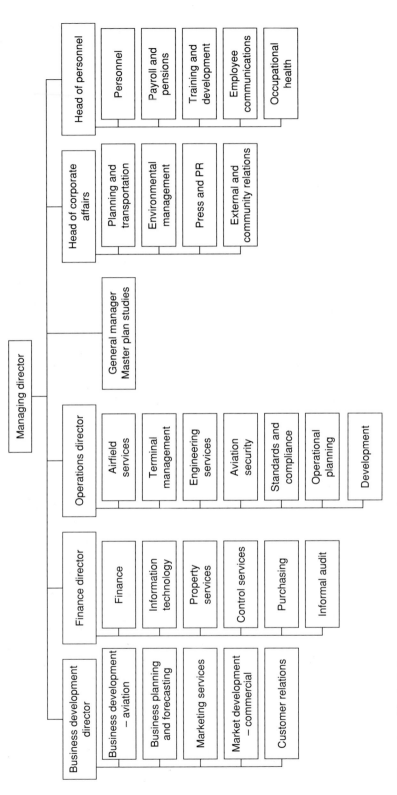

Figure 9.2 Typical airport management structure.

early-morning and late-afternoon/early-evening flights. The issue of night flying interacts with this, as neither the passenger nor the local community wants flights before 7 a.m. The operational requirement for airlines means that they want to keep their aircraft in the air on passenger-carrying journeys as much as possible. The flight journeys around Europe tend to involve about a four-hour return; thus, if an aircraft takes off at 7 a.m., it returns and takes off again in the late morning, mid-afternoon and evening. This induces capacity requirements at these times.

The primary determinant of capacity of an airfield is the runway, although taxiways and terminals and the configuration of these are important, with other local factors such as wind coverage, time and track restrictions, airspace utilisation and availability and type of navigational aids providing constraints. The runway is the centre of airport operation; its width, length, strength and direction are fixed in time. The strength, length and width determine the size and type of aircraft that can operate and whether they can operate at full load. The siting and direction of the runway determine the influence of the weather and of local restrictions, whether topographical or community-derived. Strong cross-winds can prevent landing; therefore, runways are sited to be available on average. Topographical conditions and community restrictions may prevent the use of runways in certain conditions, which severely reduces capabilities. Having another runway is not a trivial move: it is a quantum of expansion in both operational and financial terms. In particular, it involves an extended planning period, which might take 10 years, involving public enquiries and being called in by the secretary of state. Such expansions will be part of the master plan that airports maintain; the national political position on this was set out in the aforementioned government White Paper (DfT, 2003). Thus, airports search for ways to extend capabilities by other means.

All runways operating commercial aircraft have an instrument landing system that is graded by ICAO categories. This assists pilots to land in poor visibility conditions and, with the highest category, to perform automated landings. Such systems allow landing in most weather conditions, thus extending the capability of the airport. Such technology develops alongside aircraft-landing technology to further extend the operating capacity of airports.

A runway can accommodate about 50 aircraft operations an hour, although landings have priority and require more time. At peak times, the flight path to the airport may require stacking in order to buffer the arrival of aircraft. Landings need to be spaced in order to allow aircraft to safely abort their landings. In addition, aircrafts produce and trail air vortexes, which disturb following flights and may determine safe separation. Some aircraft and loadings require longer runways; however, capacity can be increased by getting aircraft to exit runways as quickly as possible. Rapid exit taxiways along the runways allow such manoeuvres and can, on average, increase capacity. Other taxiway improvements can mean that the runways are not required for taxiing, thus leaving them more time for takeoffs and landings.

Although aircraft can land and take off, the unloading and loading of passengers and luggage can become an obstacle to airport capacity. In addition to this transfer from building to plane, and vice versa, there is the movement of people through the airport. Airlines experience delays when passengers do not arrive at flights, and the early and expeditious processing of passengers is critical. Many processing features are concerned with security, and it is essential that a fully screened population and its luggage is created on the airside. The development of reliable and rapid technology for doing this is a growing feature of airports. Arrivals processing can create passenger delays, such as at customs, immigration-control points and baggage return; these do not affect the operation of the airlines, but they can clog up an airport fairly rapidly. The technology associated with baggage-handling, such that it arrives at the correct aircraft and can be loaded quickly, is critical for efficient airline operation, particularly where rapid turnarounds are required.

Although the runway is the central built feature of an airport, the terminal buildings are the gateways for, and the means for processing, passengers. The basic airport dilemma is present in the terminal building, which balances passenger processing with commercial activities. At some point, an airport terminal capacity may have to be extended, whereby new terminals will allow more passengers to flow through, and this is to be undertaken in a more efficient manner. The development of new terminals is a major undertaking and will be included in airport master plans and be subject to planning permissions. As more people are going to use the airport, extra carparking and new surface access arrangements will be required. External carparks and surface access are also part of the built environment that establishes the nature and attractiveness of the airport.

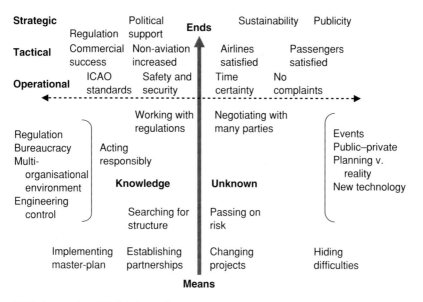

ICAO, International Civil Aviation Organisation.

Figure 9.3   Means and ends of airport development.

9.4   EXPERIENCE OF BUILDING: FROM UNKNOWNS AND CONTRADICTIONS TO MEANS AND ENDS

Airports are highly organised and have extremely rational processes, and yet they can suffer as issues beyond or contradictory to these processes transpire. Airports are complex and exist in a volatile environment, which has great opportunities and yet potential disasters. Figure 9.3 summarises the approach of airports to circumstances.

Public–private dilemma

Airports have a confusion about their constitution as to whether they are public or private enterprises, which affects the way in which they think about and experience the world. Airports are businesses but are severely restricted by regulation as to what they can do and how they can operate. The purposes of airports include a social duty, as they are public transport facilities and affect the local economy and environment. The provision of public services via private organisations is a contentious issue, particularly as many used to be in public ownership. That an airport is a public service and provides for the public good is open for debate; however, it clearly does influence the local public. The public good is demonstrated in the provision of access to transport for business and leisure and also in the general economic development in the area. This is the argument used by airports to convince local authorities to not only accept their existence but also acquiesce to future developments, particularly as local authorities control planning permissions. The national social duty for general air travel may be in conflict with a local social duty to protect its community environment. Airports have sought sources of revenue other than aviation, and thus commercial activities have grown in importance. This gives a further contradiction in objectives and a dilemma when planning future expansion.

It is evident that the purpose of an airport is confused between supplying a public service and profiting as a commercial organisation. There is even a residual military duty, which may present another stakeholder in strategic and operational negotiations. These contradictions in purpose tend to be managed by a residual bureaucratic form of operation, which reflects the airports' public-sector history and which is reinforced by the regulated world in which airports work. Bureaucracy is encouraged by the many public control authorities (police, Special Branch, customs, immigration), which have an operational duty and a tactical influence, as they require space, have a role in the processing of passengers and freight, and must be consulted in all developments on the arrangements at the airport. These control authorities operate in bureaucratic ways, both tactically in their parent organisations and operationally in the response of their staff.

Many activities span the strategic, tactical and operational levels, for example security, and this creates problems at the lower levels of authority. At the higher level, policy is declared to be implemented at the lower level; however, it may not be possible to make it happen, or there may be conflicts with other policies, which then span differentiated departments. This means that functional departments can work to their restricted purpose, and decision-making can be elongated by interdepartmental disputes. Regulation and policy, with their possibilities of different interpretations, then become tools for these conflicts of purpose.

Authority in multi-organisations

The role of the airport as a port can be compared with a dock and a town. Although the airport is a place of transit, it operates as a town. The airport is not a single organisation but a space within which many organisations operate. Airports have outsourced many functions, and thus the number of organisations involved in the airport operation is even larger. This space is owned and managed by a central company; however, the owner may not be the manager. The management company has to coordinate the various organisations to make the whole work, although many of these are under managerial contractual control, e.g. cleaning. They have managerial authority, legislative authority and procedural authority through by-laws set by Sections 63 and 64 of the Airports Act 1986 and Section 37 of the Criminal Justice Act 1982. In operational terms, the manager can stop things happening because of the manager's policing role and authority, but the manager is much less capable of making things happen. This requires negotiating with the various organisations that exist as part of the airport.

The airport–airline relationship is strange, as the airline can move to a different airport. The influence of the airlines and developing aircraft technology are always a threat. There is also a dilemma about investing in technical change. This not only involves capital equipment cost but also often requires extensive maintenance contracts and staff training, which are ongoing costs. If airlines choose to operate different aircraft that require different handling equipment, then this needs to be purchased or leased. This may also require changes to buildings in order to accommodate this equipment and new aircraft. Such investment is dependent on the success of the airline and the technology, and so the controlled world of the airport is put into a risk situation.

Environment or business

This contradiction is so invasive and so intractable that it affects all activities. Politically, there is an agenda of economic development that sees airports as an excellent opportunity to regenerate and sustain local businesses and jobs, but at the same time there is also an agenda of environmental protection that sees airports as an evil. Individuals and organisations are polarised in these views. The debate involves putting down the opposing position, e.g. 'raping the planet' (Reid, 2005) on the one hand and 'beans and lentils brigade' (Kenan, 2005) on the other. Airports have to present both positions, namely by having in place extensive environmental policies and monitoring and including within their master plans a detailed report of their potential economic benefits. This contradiction will also be evident in the local authorities surrounding the airport and indeed is present in the government's strategy (DfT, 2003). This causes lurches in decision-making, as other events affect the weight of the argument, e.g. hurricanes at the time of a decision indicate global warming and suggest the restriction of development, whereas the closure of a large factory in the area supports the need for economic development. This feeds through into current developments certainly at the commissioning stage but also during construction, when protests may occur or project incidents may be exaggerated and used for political pressure.

Although airports promote access by public transport, there is a contradiction in that a significant level of their income comes from carparking. For example, at a regional airport where 40% of income comes from commercial activities, about 20% of this (or 8% of total income) comes from carparking. In smaller airports, where retail is not so significant, carparking is a more major part of this income.

Environmental issues are important concerns because of publicity and future fluidity in negotiations. All activities are scrutinised by campaigners and thus may become the subject of attack. Contractors need to be vigilant in all projects that could give rise to pollution. Casualness that might be acceptable outside will interfere with airport tactics and yet may be required in order to make things happen.

Security or access

The airport that is running a business is well aware that security is costly and a tremendous inconvenience and invasion of privacy for its customers. Therefore, the airport has to balance total security against its ability to run a viable business. Much of this involves expensive scanning and screening equipment, which requires staffing and adequate maintenance because of the potential liability for delays. Security people may have an authoritarian view of the world that does not make them customer-friendly, which can induce more problems. The airport has an obligation to its customers with special needs that are difficult to meet. In addition, it also has to process VIPs, who have their own personal security problems and attract attention from the media and the general public. This is a risk-management exercise that will throw up different solutions as new events transpire.

Flying is emotional; it is exciting, yet frightening. Flying is still a risky activity based on the vulnerability of technology and human interaction with this. Passengers have a belief in, but also a fear of, the technology. Also, people travel to seek opportunities but at the same time are leaving current arrangements. These emotions become difficult to manage as they are in the unknown of the airport, and these can diffuse out into projects.

Rational planning versus reality

Large and regional airports have long-term views based on a pre-dicted major rise in air use. The implementation of these plans requires significant funding from a commercial world, which requires a rational business plan. The lower the risk, the less interest; thus, it is worth presenting a favourable plan. As events transpire, however, it may not be possible to meet this plan, which requires remedial measures. Uncertainties of politics, regulation and air incidents can change demand almost instantly and certainly during longer projects. The gap between the reality and what is presented can be high and can induce changes in direction, including the abandonment of projects, which may not be realised because of external factors and developing operational constraints.

Control and time are important to airports, and an ability to work within their constraints is essential. Larger airports are good payers but require flexible working and seek both value and commitment to better match developments to their growing business. The con-struction industry needs to learn to plan effectively for airports and to do this within a multi-organisational setting. The latter requires good communication skills and an ability to negotiate across boundaries. Thus, partnering, alongside other, longer-term relationships, becomes an attractive option for regional airports as a way of managing this uncertainty.

At small airports, the economics are more vulnerable to external factors and the local market. This makes them extremely cost-conscious and likely to seek lowest-cost solutions. They require extremely tight business management. As opportunities arise, e.g. an airline seeks a new route, these airports are forced to make attractive offers and to provide the necessary facilities to accommodate the developments. Again, as the airline is free to move elsewhere, this is a vulnerable investment, the returns have been reduced because of the deal done by the airline, and there is a temptation to try to transfer risk on to others, which can include the construction industry.

Any changes to capacity require extended building programmes, which need to be undertaken while the airport is operational. This condition determines that these programmes are complex and are affected by normal deviations in aeroplane operating conditions. There are multiple time horizons in airports, which need to be planned for differently. There are summer and winter cycles, with fewer flights and different destinations in the winter; there is a weekly cycle, whereby weekends have fewer flights; and there is a daily cycle, with a passive night period and daytime peaks. The latter induces a maintenance horizon, with opportunities at night; however, the runway or buildings need to be available for the next day. This demands night-time working and a requirement for methods that are rapid and yet reliable. Planning is essential, as time is critical and permissions need to be obtained. Thus, there is a degree of throwing resources at problems in order to deliver solutions.

9.5   KEY ISSUES

❏ Airports are not single entities but multi-organisations.
❏ Most work will interact with a number of organisations, at least as stakeholders if not as joint clients.
❏ The operation of airports is heavily legislated, regulated and monitored.
❏ There is conflict between the public-service role and the business economics of airports.
❏ Large airports are monopolies that are regulated.
❏ Commercial activities are as important as aviation activities in the generation of revenue.
❏ There is conflict between local economic development and environmental concerns.
❏ Airports are beholden to airlines, which negotiate for discounts and slots and can transfer to other airports.
❏ Runways are the heart of airports and limit the capacity.

❏ Efficiencies of passenger flows involve terminal position and layout.

❏ Security is paramount, requiring separation, monitoring, and policing.

❏ There are multiple time horizons in airport operations, which require different management.

❏ Airports have a master plan for strategic development.

❏ Airports need to be seen to be proactive in environmental issues.

❏ Environment and security are important at political, tactical and operational levels.

❏ Airports are rational but also emotional environments.

REFERENCES

AOA (2005) *UK Aviation Facts 2005*. London: Airport Operators Association.

BHX (1995) Vision 2005: Master Plan. Birmingham: Birmingham International Airport.

BHX (2005) Surface Access Strategy, July 2000–July 2005. Birmingham: Birmingham International Airport and the National Exhibition Centre.

Bruinsma, F., Rietveld, P. and Brons, M. (1999) Comparative Study of Hub Airports in Europe: Ticket Prices, Travel Time and Rescheduling Costs. Research memoranda. Amsterdam: Faculty of Economic Sciences and Econometrics, Vrije Universiteit. ftp://zappa.ubvu.vu.nl/19990050.pdf.

CAA (2004) Economic Regulation of Airports: General Guidance. London: Civil Aviation Authority.

CAA (2005) Economic Regulation Group Business Plan: 2006/2007–2010/2011. London: Civil Aviation Authority. www.caa.co.uk/docs/5/BusinessPlan06_10webversion.pdf. Accessed 27 March 2006.

Castells, M. (2001) Information technology and global capitalism. In *On The Edge. Living With Global Capitalism*, eds W. Hutton and A. Giddens. London: Vintage.

DfT (2002) *Airport Security: Report by Sir John Wheeler*. London: Department of Transport.

DfT (2003) *The Future of Air Transport*. London: Department for Transport.

Graham, A. (2003) *Managing Airports: An International Perspective*. Oxford: Elsevier.

Gunston, B., Bently, P., Armitage, M.J., Pyle, M. and Chemel, E. (1992) *Chronicle of Aviation*. Liberty, MO: JL International.

Kenan (2005) Shaping the Future of Air Commerce Worldwide. Chapel Hill, NC: Kenan Institute of Private Enterprise, University of North Carolina. www.sovereign-publications.com/kenaninstitute.htm. Accessed 27 September 2005.

Reid, M. (2005) How our planet is being raped by cheap air travel. *The Herald*, 13 September 2005.

RESOURCES

CAA (2004) *Airport Regulation: Looking to the Future – Learning from the Past*. London: Civil Aviation Authority.

Airport Operators Association (AOA): www.aoa.org.uk. Trade association that represents the interests of 71 British airports, comprising all of the nation's international hub and major regional airports in addition to many serving community, business and leisure aviation.

Airports Council International (ACI): www.aci-europe.org. Worldwide professional association of airport operators, headquartered in Geneva, representing over 1,530 airports in 175 countries and territories worldwide.

British Air Transport Association (BATA): www.bata.uk.com. Represents vast majority of UK airlines; members produce over 90% of UK airline output.

Civil Aviation Authority (CAA): www.caa.co.uk. UK's independent aviation regulator, with all civil aviation regulatory functions (economic regulation, airspace policy, safety regulation, consumer protection) integrated within a single specialist body.

International Air Transport Association (IATA): www.iata.org. Prime organisation for inter-airline cooperation in promoting safe, reliable, secure and economical air services for the benefit of the world's consumers.

International Civil Aviation Organisation (ICAO): www.icao.int. Specialised agency of the United Nations linked to the Economic and Social Council (ECOSOC) that operates the Convention on International Civil Aviation, also known as the Chicago Convention.

10 Housing Associations as Clients

10.1 INTRODUCTION

The provision of social housing is a necessary condition for a stable post-industrial society such as Britain. Social housing supports the economy through sheltering and locating the workforce and ensuring that socially disabling activities, caused by disadvantages in a competitive society, are minimised. There is both a moral and a political element in this, surrounding the nature of the society we want to live in. Currently there is tremendous social pressure on people to own their own houses as a symbol of independence, status and worth. Houses, as property, are no longer seen only as accommodation but are increasingly seen as an investment by individuals. This has elevated the cost of housing and made it difficult for many people to buy. It has also elevated the cost of renting property, particularly in areas of high economic activity that draw in people to work.

Social housing is targeted mainly at people who need affordable rented or low-cost home-ownership options. Registered social landlords (RSLs) such as housing associations are becoming the main providers of social housing, both for rent and for sale and both for new-build and for refurbished accommodation. The administration of social housing is organised for the government by the Housing Corporation. The Housing Corporation is a non-departmental governmental body sponsored by the Office of the Deputy Prime Minister (ODPM) and is responsible for registering and regulating RSLs in England. The Housing Corporation funds RSLs through the Social Housing Grant. Similar roles are performed by the Northern Ireland Housing Executive, Communities Scotland and Cymraeg (Housing Directorate in Wales). The Housing Corporation has extended the remit of RSLs to accommodate key workers such as nurses and other hospital staff, because of the high cost of housing, and other disadvantaged groups, such as elderly people, disabled people and people from ethnic minorities, in order to remove the burden from social services.

RSLs consist of housing associations in the main, but they also include trusts, cooperatives and private companies. Most RSLs are registered charities operating as not-for-profit businesses, and thus their surpluses are used in either maintaining existing homes or building new homes. There are over 2,000 RSLs in England, currently managing around 1.5 million homes. Many of these RSLs are small and own only a few homes. Some, however, are much larger: about a dozen RSLs own and manage over 20,000 homes each (see Table 10.1).

Many RSLs have their origins in the nineteenth century or earlier, but the movement can trace its growth to the 1970s, when the Housing Corporation was given the remit to provide housing association grants (HAGs) in order to encourage development. At this time, there was a strong political move away from state-provided services and a move towards greater owner occupation. The latter created the right-to-buy legislation, whereby local authority tenants could purchase the properties they were renting at discounted prices. Following the Local Government and Housing Act 1989, housing stock was transferred from local authorities to RSLs, leading to the formation of many new RSLs in order to manage these. Malpass and Mullins (2002) estimate that almost 600,000 homes were transferred to RSLs in England within the period 1988–2001, and it remains government policy to transfer stock from local authorities to RSLs. However, a majority of tenants must vote for transfer in order for this to take place, and in some areas (e.g. Birmingham City Council, Dudley Metropolitan Borough Council, Bromsgrove District Council, Cannock Chase District Council) tenants voted against transfer. Therefore, many local authorities still manage houses for the same target group as RSLs. The rental levels for a RSL house are perceived to be higher than those for a local authority house but below those for private renting, although a process of rent restructuring is intended to even out such differences between RSL and local authority rent levels by 2012.

The ODPM is driving the housing agenda in England and wants every home to achieve the Decent Homes Standard by 2010, a 'decent home' being one that is reasonably warm and waterproofed, has modern facilities and services, is in a reasonable state of repair and meets

Table 10.1 Largest registered social landlords (RSLs) according to stock owned and managed.

Rank	RSL	Total stock		
		Owned	Managed	Owned and managed
1	North British Housing Limited	46,337	346	46,683
2	Sanctuary Housing Association	30,665	4,900	35,565
3	Anchor Trust	34,895	23	34,918
4	Hyde Housing Association Limited	17,921	14,822	32,743
5	Home Group Limited	31,550	–	31,550
6	London & Quadrant Housing Trust	25,533	1,237	26,770
7	Home Housing Association Limited	–	25,056	25,056
8	Riverside Housing	22,504	643	23,147
9	Orbit Housing Association	16,547	5,257	21,804
10	The Guinness Trust	20,298	915	21,213
11	Northern Counties Housing Association Limited	12,828	6,632	19,460
12	Circle Thirty Three Housing Trust Limited	493	17,977	18,470

The data in this table are based on the Regulatory and Statistical Return (RSR) Survey for the year ended 31 March 2004.

Source: Housing Corporation.

the current statutory minimum standard for housing. In 1997, 2,100,000 houses were owned by local authorities and housing associations that did not meet the Decent Homes Standard. Local authorities were required by government to complete a stock options appraisal, and one objective was to determine whether stock transfer to RSLs was the best approach to achieve the Decent Homes Standard. Strategies outlined for local authorities to achieve the Decent Homes target are threefold: the use of arms length management organisation (ALMO) and private finance initiative (PFI) and transfer of stock to RSLs.

10.2   BUSINESS ENVIRONMENT OF HOUSING ASSOCIATIONS

Although housing associations are private companies, the social service delivery ethos drives housing association organisations; thus, domain theory (see Section 3.3) is the most appropriate model for us to understand the way in which a housing association delivers its objectives to its environment and receives resources from its environment. Domain theory distinguishes both a different purpose and a different set of achievements at political, managerial and professional/ operation levels. The business environment of a housing association is presented in Figure 10.1. This is generalised, and local differences will exist. We will describe the political domain, which involves government policy and regulation, followed by the managerial domain, which involves regeneration and development, and then the operational domain, which involves the community and service to tenants where many of the political issues face reality.

National political domain

The provision of housing is a political issue that has been high on the agenda for over a century. Housing is one of the current big issues of concern in the UK, because house prices have risen sharply. Many first-time buyers cannot afford a house of their choice, and lower-income people find it difficult to rent a reasonably decent home. Like all social service issues, there are conflicts between what the state should provide and what individuals should provide for themselves. Housing policy is an electoral battle ground, and the successes and failures of governments are exposed in partisan ways. Housing can

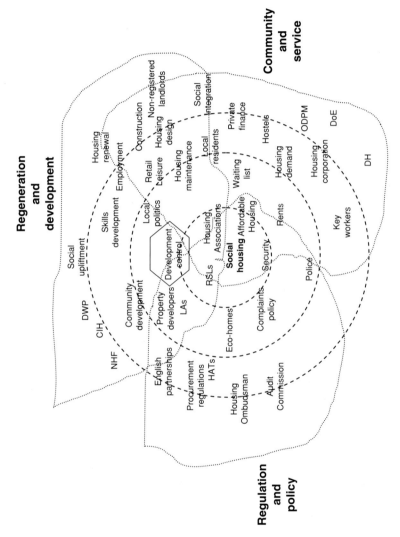

CIH, Chartered Instituted of Housing; DH, Department of Health; DoE, Department of Environment; DWP, Department for Work and Pensions; HAT, Housing Action Trust; LA, local authority; NHF, National Housing Federation; ODPM, Office of the Deputy Prime Minister; RSL, registered social landlord;

Figure 10.1 Map of social housing sector.

require substantive resources from the Treasury, so that there is competition with other government activities. Change in housing provision, however, can occur only in the long term, and so there is a limit to electoral benefits to governments.

The government creates agendas that the housing associations are obliged to address, although they might not follow them to the letter. The government's key agenda is the concept of sustainability, which involves meeting the economic, social and environmental needs of future generations. Sustainability is tied in with the regeneration of communities. RSLs do not only provide homes but also develop communities. According to Wilson (2003), the real issues of quality lie outside the home, i.e. in the neighbourhood. There is every need to regenerate communities because if good houses are provided in deprived areas and the socioeconomic standards of these areas are not improved, then the houses will degrade quickly. There is no point in providing decent houses where crime and vice would thrive. Thus, all aspects that contribute to making an area sustainable need to be considered in developing new social housing, for example community safety, education and childcare facilities, transport links, health services and leisure facilities.

In February 2003, the ODPM launched the Communities Plan (Sustainable Communities: Building for the Future), setting out a long-term programme of action for delivering sustainable communities in both urban and rural areas. This programme was spread over four years (2002–06) and was worth £22bn. The plan included addressing housing shortages, especially in the north/south-east of England; low and changing demand of housing, especially in the north-west and the north Midlands; the Decent Homes Standard for all social housing by the year 2010; improvements to the local environment of all communities, for example cleaner streets, improved parks and better quality of public spaces; and protection of the countryside. Government priorities currently include spending on housing for key workers and building in London and the south-east, along what is known as the Thames Gateway.

In order to maximise efficiency of provision and reduce its administrative burden, the government wants to rationalise the social housing sector by reducing the number of RSLs that it funds. The Housing Corporation is thus guiding RSLs towards mergers, group structures and clusters around lead-development partnerships. Thus, large housing groups are emerging, and these are entering into partnership arrangements with several smaller RSLs. The opportunity to form new RSLs that would be viable is encouraged only in specialist areas, for example developing minority ethnic housing associations.

The social housing sector is highly regulated, with emphasis on the provision of decent standards, effective services and fair treatment of tenants (Mullins *et al.*, 2001). The Housing and Town Planning Act of 1919 required all local authorities to provide publicly funded housing in order to meet the demand in their areas. This continued for about 60 years, until 1979, when the rolling back of the welfare state began to gather pace, which affected the housing sector much more than other sectors (Franklin, 1998). The Decent Homes standard set by the government in 2000 requires all properties by 2010 to be free from dampness, be watertight, have kitchens that are less than 20 years old and have bathrooms that are less than 30 years old. Each property needs to be evaluated on a risk scale, based on the most vulnerable person inside. Thus, the regulation goes into the detail of the operations of the organisation. Other regulation surrounds how the housing associations undertake their business, which requires housing associations to demonstrate best value. In the construction arena, this now also includes that housing associations become best-practice clients. The regulatory code requires all RSLs registered with the Housing Corporation to have a complaints policy and be a member of the Independent Housing Ombudsman Service. The Housing Act 2004 aimed to consolidate a fairer housing market for all those who own, rent or let residential property and to protect the most vulnerable people in society by building community cohesion and recognising diversity.

In June 2005, the Housing Corporation announced that future affordable homes ('Eco-Homes') should reduce carbon dioxide emissions by 26% compared with the average of current new-build houses. This involves the new-build Eco-Homes achieving 60 out of 100 'Eco-points' compared with the current average achievement rate of 36 points under the Building Regulations of 2002.

A combination of government funding and other source(s) is used to finance social housing. One of the other sources of funding is the trading surpluses of RSLs. Rents from tenants accrue to RSLs. Also, where there have been stock transfers, transferred tenants retain the right to buy (RTB); the RSLs retain some of the capital receipts from such sales, and these can be reinvested in housing. RSLs are also able to raise finance through loans and mortgages from banks and building societies, servicing this through the revenue they generate. The stock transfer from local authority to RSL allows an RSL to use the difference between the exchange value of the property and its existing debt to borrow on the equity and invest the funds to improve its service (Garnett, 2000). Finally, the PFI is available for social housing projects, following a number of first-wave pathfinder schemes in this sector in March 2000 (Grubnic and Hodges, 2003).

Competition in social housing provision is between RSLs because they have to bid for funds from the Housing Corporation. RSLs have to justify need, affordability and other criteria in order to secure funding. The balance-sheet capacity to deliver larger volumes of programmes is now guiding Housing Corporation funding. The government wants a step change to reduce the number of providers, so it can control this better in terms of larger programmes over longer periods of time. In order to fund these, RSLs must have the balance-sheet capacity that can deliver the development, with private-sector borrowing supplemented by Housing Corporation funding, which requires the RSLs to have the asset base to facilitate such borrowing from the private sector. RSLs must also have development expertise and a track record that shows they can deliver homes efficiently. In the past, the Housing Corporation funded only RSLs for housing development. Now, through the Housing Act 2004, non-RSLs are entitled to funding from the Housing Corporation. Thus, there is potential competition from private-sector non-registered but socially focused landlords, who can now bid for funding, compounding the competition.

Local political domain

Although national politics sets much of housing associations' purpose and means of operation, housing associations also have to work within a local political arena. In order for RSLs to receive development grants from the Housing Corporation, they need to work in partnership with local authorities to develop new communities and sustain communities. Before housing solutions are devised, the local community and (potential) end users need to be consulted in order to determine precisely what they want or need. The essence is for the community to feel the impact of a scheme, be it modern infrastructure, social upliftment, acceptable service standards or sustainable environment. There can be conflict, however, with local authorities because of their stock transfer, which often had a backlog of repairs and improvement works. RSLs also need assistance from local authorities with regard to the availability of land and the ability to obtain planning permission.

Also introduced by the government in 2002 was the rent-restructuring scheme for affordable housing, wherein all landlords operating similar types of property in the same locality have to charge an equivalent level of rent. For any new development, an RSL has to indicate the start rents when applying for development funding. Afterwards, the RSL cannot change those start rents and later changes will then be controlled by increases not exceeding RPI plus 0.5%.

RSLs need to liaise with agencies such as primary care trusts and with retail sectors, as new developments are required to include community facilities such as surgeries and shops within their development schemes. Housing associations are also able to provide jobs through initiatives such as New Deal for Communities by employing local contractors or even in partnership providing nursery places so local families can re-enter the job market.

Local private housing developers are interested in working with housing associations because all new housing developments now have to include 15–50% of affordable housing in their schemes, as part of planning requirements. It is difficult to determine, however, what is meant by 'affordable', as there are no hard and fast definitions of this terminology. This means that joint ventures are possible with private developers that provide social housing; however, this requires strong negotiating in order to obtain the maximum benefit to the social housing side.

10.3 MANAGEMENT OF HOUSING ASSOCIATIONS

Politics shapes the services and products that RSLs can provide in a strategic way. The management of public service organisations has to mediate between the operational difficulties and these political structures. In a housing association, which is a private company, there is the additional pressure of accommodating a strategic purpose with long-term viability. This is set within the expectations of a public body in the sense that it needs to be transparent, consultative and accountable. Housing associations achieve this through having a set of policies, procedures and open-decision committees. The desire for building comes from this domain, but the use of buildings is in the operational domain.

Many housing associations were formed as small organisations in the 1970s and have metamorphosed into large group associations through strategic mergers, acquisitions, stock transfers and organic expansion. The group structure has a parent company with several subsidiaries, which represent regions or special activities. Some of these group companies can be charities, while others can be constituted as private development companies. Large housing associations may turn over £160m a year and have assets of £1.5bn, and their stock value is ever on the increase.

The business of an RSL is overseen by a board involving directors, trustees or governors, many of whom are volunteers and will include tenants, representatives from local authorities and community groups, business-people and local politicians. There will be an executive management group responsible for the day-to-day operations and advising the board about strategic direction. Initially empowered by Section 36 of the Housing Act 1996, the Housing Corporation launched its Regulatory Code in April 2002, which requires RSLs to demonstrate that their organisations are viable, properly governed and well managed. An overall management structure for a typical housing association is shown in Figure 10.2. There will be an asset manager, whose responsibility is to ascertain the condition of the stock and the demand for different types, in order to identify what is needed. A development division will be responsible for developing new housing schemes and for checking whether a proposed scheme is sustainable before the executive board approves this. In a group structure, it is normally the regionally based subsidiaries that identify the projects that are viable because they are in touch with the communities where they operate.

There is no direct political involvement in the day-to-day activities of RSLs. The Housing Corporation will intervene only when an RSL fails in its performance. The Housing Corporation will then use its regulatory powers to bring the failing organisation back on track. This can include close supervision of the running of the organisation, replacing board members, replacing senior executives and transferring the stock to another RSL. Therefore, RSLs operate in a regulated environment, but direct political intervention in their affairs is almost non-existent compared with the influence of both central and local politics in local authority housing provision and management. The Housing Corporation monitors the regulatory performance of each RSL annually on four standards: governance, financial viability, scheme development and management. A traffic-light system is used for this purpose, whereby

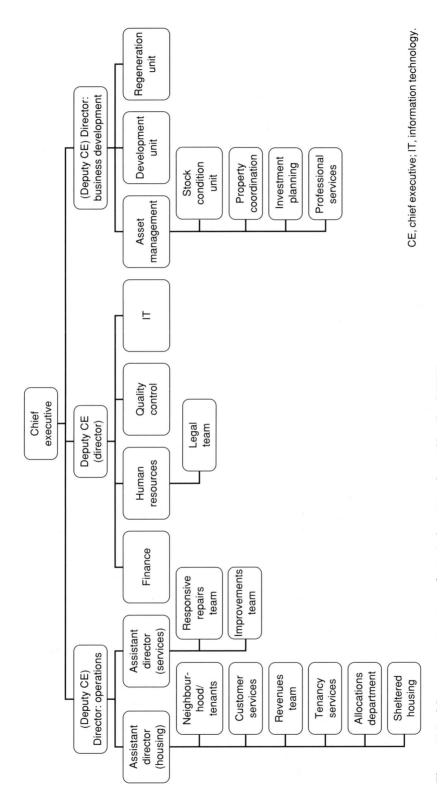

Figure 10.2 Management structure of a typical registered social landlord (RSL).

CE, chief executive; IT, information technology.

an RSL can hit red, amber or green on any of the assessment criteria. The outcome of the Housing Corporation's assessment is published publicly (see www.housingcorp.gov.uk). The other external way in which RSLs are monitored is through the biennial inspections of the Audit Commission, which monitors performance concerning repairs and maintenance, tenant involvement, equality and diversity, lettings, antisocial behaviour and customer service. RSLs are rated by the Audit Commission as one, two or three stars, with three stars being excellent, two stars good and one star fair. There are subcategories within each rating. Although this rating system is set to change, this will involve transferring responsibility or developing new methods rather than a new objective. Like the Housing Corporation, the Audit Commission publishes its inspection results (see www.audit-commission.gov.uk).

Most RSLs belong to the National Housing Federation, which promotes the work of RSLs and negotiates with government. The Chartered Institute of Housing is the professional body for people working in housing. These two professional organisations help lead the way in promoting best practice in the social housing sector.

Viable business and conflict of purpose

As businesses, housing associations seek to be market leaders in their areas in terms of quantity, quality, innovation and performance of their portfolio. Their end products are varied and include property for renting at market rates, housing for key workers, commercial housing for sale, student accommodation and care homes. Many properties are on estates, but some are 'pepper-potted', whereby there are only one or two properties on a street or in an area. This mix makes equality of service provision more challenging. Most housing associations believe in growth in order to use their resources better to provide more affordable homes and improved services. The spread of their businesses can be wide in order to achieve growth, going wherever government priority and funding is and wherever there is a stock-transfer opportunity.

An RSL will operate, say, a five-year business plan, which is reviewed year on year. This plan sets the number of houses that the RSL will develop every year, the staff to employ, etc. Internally, the RSL monitors its performance through internal audit and by benchmarking against other RSLs. Processes are reviewed regularly to ensure efficiency and use of best-practice techniques. Demand projections underpin stock-transfer acquisitions, investing only where the demand potential is high. Demand also controls where RSLs build. For many years, the Housing Corporation zoned housing association development activity, and this meant that many RSLs developed in specific geographical locations, with only a few developing on a national or multiregional scale. Projects are planned and priced in advance. Before a funding bid is finalised, a lifecycle-costing and cash-flow model forecasts the financial performance of the scheme over a 30-year period. It is deemed that after 30 years the scheme would require a major regeneration, and the Housing Corporation accepts this forecast period.

Although housing associations are attempting to regenerate communities, the buildings they develop need to be viable businesses. The right quality of house, in the right environment and with the right aesthetics is required. Another tactic is to incorporate commercial facilities in schemes, including shops and doctors' surgeries, which can bring in additional revenue. Additional facilities, such as retail, leisure, play areas, parks, parking space, rivers, ponds, transport, schools and crèches are viable where developments are of a scale of several hundreds or a thousand units. A comprehensive solution that befits the area is often considered, taking due cognisance of what the community wants. A tactic used to sustain communities is to mix together houses that are rented and houses purchased through shared or sole ownership. These houses will appear similar on the outside, but internal fixtures and fittings might differ. In this way, the exterior is coherent and residents are not stigmatised.

A central debate in housing management has been whether it should adopt a purely property-management role or a more welfare-oriented role (Saugeres, 1999). The provision of welfare, like social housing, is a scarce resource within the welfare state that can be allocated to only a limited proportion of the population (Klein *et al.*, 1996). The social housing sector has a business dilemma as it caters for the poorest sections of the population who cannot afford any other form of housing, but who may not make good paying tenants (Cole & Furbey, 1994; Malpass & Murie, 1994). New community-care legislation encourages people who formerly would have been in institutions to live in their own homes in the community (Franklin, 1998). Thus, many RSLs are developing homes and schemes for elderly people and disabled people. RSLs now develop 'lifetime homes' on some projects, which are habitable by people through the physical deterioration within their life (Carroll *et al.*, 1999). Some characteristics of these lifetime homes are wider doors that can be adapted for disabled residents and specially designed roof trusses that allow adaptation into a loft. Occupants can thus live in these houses for their entire lives. RSLs have to consider a wide spread of needs, making their task more complex than that of, for example, a speculative housing developer. RSLs also have to interact with diverse sectors in order to achieve their aims.

A further conflict of purpose can arise from a housing association's strategy for regenerating communities through developing skills and reducing unemployment. In some of the most deprived areas in the UK, it is possible to create construction training opportunities within social housing developments. In these areas, the contractors are encouraged to engage local labour in the community and to incorporate the training of new skills in the locality, giving cognisance to local demographic mixtures while not disadvantaging others unnecessarily.

When RSLs merge, there can be conflicts because the two or more companies that come together experience an entire organisational change in the way of doing things. The transition involved has to be managed culturally and procedurally. In addition, when RSLs acquire housing stock from local authorities, they are required to accept transfer of undertakings protection of employment (TUPE)-protected staff into their employment. These TUPE-protected staff may have been used to working in a more secure environment, and adapting to a more market-oriented business might be challenging to them. While many members of staff may be flexible and willing to adapt, there will be some who are comfortable with the old ways of doings things and hesitant to change and who need to be managed.

Building values

The UK government's agenda and initiatives have amplified the opportunities in this sector for refurbishment, maintenance and new-build works. Most housing associations procure development with external consultants and contractors. The whole development process is elongated because of the number of steps that have to be undertaken and the bureaucracy that is required. Within the housing association development department, there will be a team that facilitates each scheme, getting the grant, assembling the site, acquiring planning, commissioning the design, supervising the construction and handing over the development to operations. Working with a housing association requires considerable understanding of its ways of operation and how this is influenced by the government's agenda.

Designers play a significant role in social housing, as there is increasing pressure to evolve schemes that are befitting, appealing and cost-effective. Designs have an impact on the management and maintenance of housing, and this can be significant in respect of levels of crime and antisocial activity (Poyner & Webb, 1991). Designs can also promote sociability and feelings of belonging to the wider neighbourhood (Hillier & Hanson, 1989). An architect prepares a scheme design that fits with the local community. This is priced using external cost consultants and then put through a financial appraisal to see whether it will work within the grant parameters; if so, then a grant application is developed. The housing association then bids for funds from the Housing Corporation. If successful, the housing association buys the site, obtains planning permission and proceeds to the construction phase. Projects are different, and some may not follow this template rigidly. There are, however, several interfaces and likely problems along the way of each project, which can slow down the whole process.

Due to the nature of their remit, RSLs develop fit-for-purpose schemes, each project being planned to meet the needs of a particular community. However, RSL schemes are increasingly becoming innovative in design, aesthetics and construction. For example, the Peabody Trust undertook an innovative and sustainable residential building project in Surrey called BedZED. This is a mixed-tenure zero-energy development scheme. The Peabody Trust has also pioneered modular construction, with a series of London apartment blocks (Lane, 2003).

WAVE Homes is innovative development, creating timber-framed, factory-built, and eco-friendly affordable housing and offering new householders significant energy savings. The scheme is a joint public–private venture between the Riverside Group, an RSL, and Avebury, a leading housing consultancy in England. The first set of WAVE Homes, located in Milton Keynes, was handed over to BPHA, another RSL, for management in March 2005 (Riverside Group, 2005).

European Union (EU) directives influence procurement in the social housing sector. With regard to construction, many RSLs are adopting framework agreements, whereby they enter into partnership with contractors for periods of up to four years. This can be done under EU regulations, but such agreements need to lapse after four years and subsequent projects must be opened up to another phase of competition. The housing framework agreements tend to be operated along regional lines, giving the successful contractors reasonable volumes of work in each region. The procurement of all types of product is overseen by the director of development of the RSL. There is encouragement through the government's desire for best practice in the industry to ask contractors to manage their supply chains.

Thus, key to the management of housing associations is the resolution of these externally driven conflicts of purpose and with a sustainable strategic approach. These conflicts of purpose penetrate into operations.

10.4   OPERATIONS IN HOUSING ASSOCIATIONS

The operations of housing associations are dominated by providing services to customers and sorting out their problems within an environment of policy, procedures and accountability. Most housing associations have a customer panel that meets as a forum for residents to put forward their views on the range, type and quality of services provided. Residents are also given an opportunity to express their views on the priorities and proposals for changing or improving services. There will also be in place a grievance procedure, whereby a tenant can lodge a complaint against a repair person or staff member if dissatisfied with a given service or treatment. Some of the housing management functions are problematic and emotionally challenging, such as the enforcement of tenancy conditions, intervention in cases of nuisance and neighbour disputes, community development, financial advice and counselling, and care and support (Franklin, 1998).

Buildings are seen to give families a decent place to live and fulfil their aspirations; however, buildings need to be maintained on a continuous basis in order to keep the occupants satisfied. All housing associations offer a full repairing and maintenance service, although this may be outsourced. Repair problems are normally classified by the nature of attention needed, i.e.

❑ *Emergency:* requiring attention within one day.
❑ *Urgent:* requiring attention within one to five days.
❑ *Routine:* requiring attention within five to ten days.

In-house teams or contractors conduct repairs. Each time a repair is effected, the tenant involved is given the chance to give feedback on the work. Tenants and residents are encouraged to be involved in the management of their new communities; however, this can delay and disrupt plans.

With occupancy, there could be a high turnover rate, and any RSL needs to consider how to keep houses occupied. There is often a particularly higher turnover rate with younger residents, due mainly to divorce, change of jobs, change of location, etc. 'Voids' (empty or untenanted properties) are a key performance indicator (KPI) in the housing sector. Health is another area where the sustainability of communities can be developed. Health and education standards are likely to improve where houses that previously were damp and had insufficient room for families with children are upgraded to Decent Home Standard so that they are damp-free and have sufficient room for the children to do their homework. Housing associations monitor the performance of an area, looking at benchmarks such as educational achievements and mortality rates over a period of time. If such improvements can be repeated on a large scale, then the impact on society in terms of minimising crime and improving health, education, infrastructure, etc. is potentially large.

In order to develop sustainable communities, housing associations may attempt to mix the composition of occupants on an estate, i.e. with people from different backgrounds, income levels, etc. The hope is that if some people have a stake in the community and pride in its appearance, then all residents may be similarly motivated. In that way, members of a community support each other. However, social misbehaviour by tenants needs to be managed, particularly where it involves damage to property. Housing associations are becoming stricter and will evict tenants in extreme situations, but this involves considerable legal and emotional work. Ultimately, the local authority has responsibility to house evicted residents and may work in partnership with RSLs to sort out the problem, but this can be a source of conflict. There are other potential issues that can affect social housing estates, for example a high rate of theft and vandalism in an area. There may be attempts to design out or minimise such issues, for example by controlling the disposition of houses, improving the security standard of doors, windows and bolts, and using effective lighting, closed-circuit television (CCTV) and alarm systems; however, many of these problems can seem intractable on a day-to-day operational basis.

10.5   EXPERIENCE OF BUILDING: FROM UNKNOWNS AND CONTRADICTIONS TO MEANS AND ENDS

The constitution of RSLs as not-for-profit private companies sets them up with contradictions. Their agenda, however, is to provide a social service, and thus they have a public service position where their management mediates between politics and a professional service. The emphasis on policies and procedures, the filling of forms, and the classification of clients according to predefined categories that characterises all bureaucracies is prominent in social housing (Saugeres, 1999). In this regard, RSLs are bureaucratic in order to face these conflicts and contradictions. Figure 10.3 presents how housing associations' thinking determines the means and ends of development.

Although RSLs are private entities, they obtain funding from a government agency. They also liaise with other public service organisations (police force, Department of Transport, local authorities, etc.), and so they are always accessing a bureaucratic world. Although RSLs may not want to be bureaucratic, they have to be in order to be accountable in meeting government targets and thus continuing to receive funding. RSLs have to balance the contradiction between being a bureaucratic social service and running a business. Frontline housing management has to interpret policies and procedures at that point in light of an applicant's or user's specific circumstances. This can be problematic, as staff members need to be sensitive to the people they are dealing with and yet be remote while assessing and classifying their suitability. The stereotyping of residents is evident when they are denied services that people with more 'acceptable' addresses take for granted, for example the blanket redlining imposed by credit companies, rejection by employers and the reluctance of the police to enter the area (Franklin, 1998).

As RSLs work in a regulated and political environment, their processes, like funding applications, have to accord with specified criteria and be rational. However, the contrasting perspective is that regulations harbour uncertainty because they could be renewed or re-interpreted. So, an inherent risk that RSLs face is, say, government changing its priority from one region to another. Obtaining planning permission is also a risk, because it can take a long time to acquire planning permission. A fallout risk of this delay is the risk of price

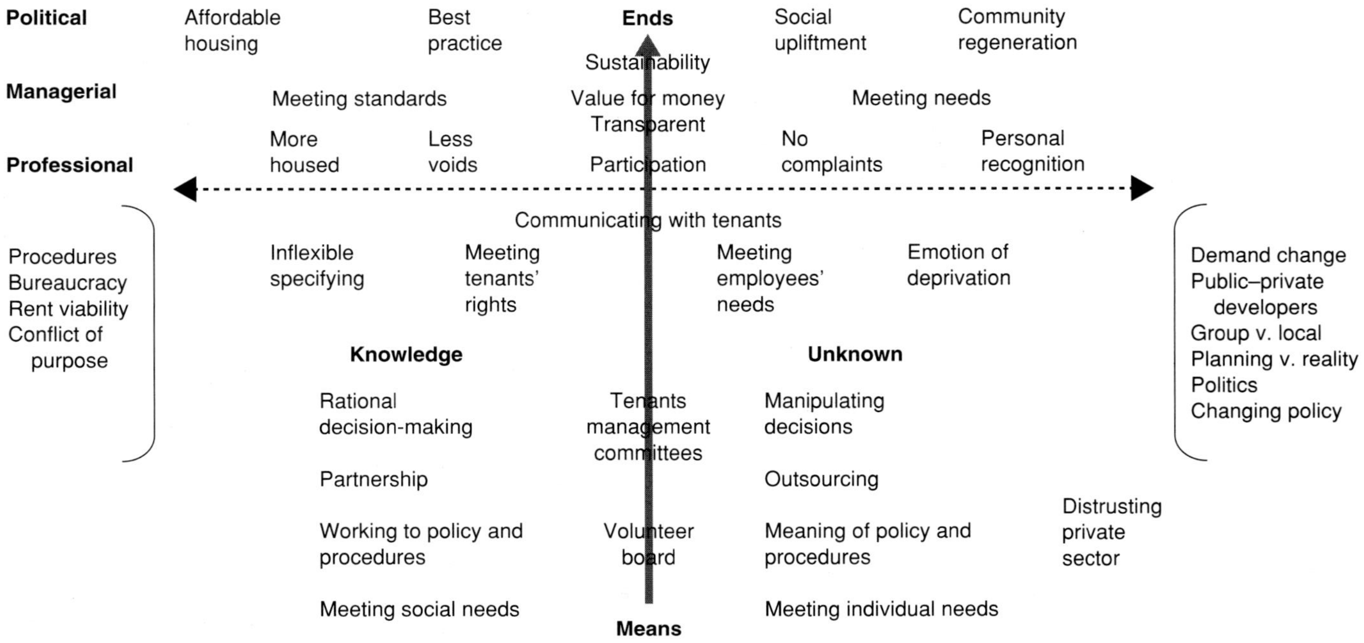

Political

Managerial

Professional

| Political | Affordable housing | Best practice | | Sustainability | Social upliftment | Community regeneration |

Meeting standards

Meeting needs

More housed | Less voids | | Value for money | Transparent | No complaints | Personal recognition

Inflexible specifying

Meeting tenants' rights

Participation

Communicating with tenants

Emotion of deprivation

Ends

Knowledge

Unknown

Rational decision-making

Meeting employees' needs

Manipulating decisions

Partnership

Tenants management committees

Outsourcing

Working to policy and procedures

Volunteer board

Meaning of policy and procedures

Distrusting private sector

Meeting social needs

Meeting individual needs

Means

Procedures
Bureaucracy
Rent viability
Conflict of purpose

Demand change
Public–private developers
Group v. local
Planning v. reality
Politics
Changing policy

Figure 10.3 Means and ends of housing associations.

increases. If interest rates go up during this waiting period, then the cost of production consequently will go up. Construction costs are a risk too: there is an acknowledged skills shortage in the construction industry, which means that contractors have to pay high wages for skilled craftspeople. However, RSLs sign to deliver a scheme at a certain price and the Housing Corporation will not increase their grant; thus, if this price goes up during the elongated processes of development, then the RSL will have to bear the impact of that risk. For a £100m project, a 5% increase (£5m) is significant. This is an area that generates and compounds emotions.

Risks and unknowns are triggered by many factors. For instance, demand for (social) housing is not consistent across the country. While there is high demand in cities such as Leeds, York and Harrogate, there is a contrasting very low demand in places such as Doncaster, Sheffield and Rotherham, where local authority and RSL properties have had to be demolished in their thousands. There is a great pressure from communities to sustain social housing, even if the demand is not there; however, the business case for housing associations cannot accommodate this, and they must be aware of the continuing demand for anything they build. From a capital point of view, contingencies are used to cover for construction eventualities, for example unknowns in the ground, delays due to bad weather, changes in specifications and builder liquidation. Depending on the risk, there could be a policy dealing with its mitigation. Some RSLs have a risk and financial management division within their group, with sophisticated risk-management principles and procedures in place to ensure that risks are managed.

The bureaucracy within RSLs extends into standardising people and problems into categories determined by written policies and procedures. This rationalistic treatment is seen as being fair to everybody (Saugeres, 1999). Rationality extends to other aspects of social housing. For instance, some RSLs utilise a strict design and technical brief in order to address the government's scheme design standards, thus constraining designers and contractors.

RSLs build to meet the aspirations of communities. However, many people aspire to own their houses and not to rent or live in social housing. Another contradiction in social housing is its suggestion that all users should consume the same standard of housing, regardless of their status, preference or ability to pay (Franklin, 1998). RSLs cannot meet some of the aspirations of tenants; if they did, their schemes would become unaffordable to both the Housing Corporation and their tenant/buyer market. So, RSLs have to balance between what is possible within the funding constraint and what is wanted by their customers. There is always a conflict with providing houses that are attractive for sale, under RTB, as this loses good stock and fragments estates. Some housing associations welcome the RTB option, as they are supporting the development of people on lower incomes. It might even be argued that some social housing communities have remained sustainable due to RTB.

10.6   KEY ISSUES

We can summarise developments in the social housing sector with these key points:

- ❏ The ultimate goal of RSLs is now to develop and sustain communities.
- ❏ Housing associations have a strategic conflict of purpose driven by changing political policy, which instils an ongoing uncertainty in their management.
- ❏ Housing associations are becoming housing developers as much as landlords.
- ❏ This sector is highly regulated and bureaucratic because of government policy and funding requirements.
- ❏ RSLs are merging into fewer, larger groups with more procurement power and greater experience.
- ❏ Buildings need to be developed in conjunction with communities.
- ❏ Housing associations require open and consultative decision-making.
- ❏ Social housing development can be slow because of the number of steps to be passed in a serial manner.
- ❏ Construction procurement is being streamlined in social housing, moving towards four-year cyclical framework agreements.
- ❏ The immediate future will witness the increasing use of supply chains for executing social housing projects.
- ❏ There is an endeavour by housing associations to identify risks in advance and to put in place strategies to mitigate these risks.

REFERENCES

Carroll, C., Cowans, J. and Darton, D. (eds) (1999) *Meeting Part M and Designing Lifetime Homes*. York: Joseph Rowntree Foundation.

Cole, I. and Furbey, R. (1994) *The Eclipse of Council Housing*. London: Routledge.

Franklin, B.J. (1998) Constructing a service: context and discourse in housing management. *Housing Studies*, **13**(2), 201–216.

Garnett, D. (2000) *Housing Finance*. Coventry: Chartered Institute of Housing.

Grubnic, S. and Hodges, R. (2003) Information, trust and the private finance initiative in social housing. *Public Money and Management*, **23**(3), 177–184.

Hillier, B. and Hanson, J. (1989) *The Social Logic of Space* (revised edn). Cambridge: Cambridge University Press.

Klein, R., Day, P. and Redmayne, S. (1996) *Managing Scarcity, Priority Setting and Rationing in the National Health Service*. Buckingham: Open University Press.

Lane, T. (2003) Bravo to zero. *Homes: The Building Housing Supplement*, **CCLXVIII**(24), 10–14.

Malpass, P. and Mullins, D. (2002), Local authority housing stock transfer in the UK: from local initiative to national policy. *Housing Studies*, **17**(4), 673–686.

Malpass, P. and Murie, A. (1994) *Housing Policy and Practice*, 4th edn. London: Macmillan.

Mullins, D., Reid, B. and Walker, R.M. (2001) Mordernization and change in social housing: the case for an organizational perspective. *Public Administration*, **79**(3), 599–623.

Poyner, B. and Webb, B. (1991) *Crime Free Housing*. Oxford: Butterworth-Heinemann.

Riverside Group (2005) Making headlines. Internal bulletin, Riverside Group, January 2005.

Saugeres, L. (1999) The social construction of housing management discourse: objectivity, rationality and everyday practice. *Housing Theory and Society*, **16**(3), 93–105.

Wilson, W. (2003) Delivering the Decent Homes Standard: Social Landlords' Options and Progress. Social Policy Section research paper 03/65.

RESOURCES

Marsh, A. (2004) The inexorable rise of the rational consumer? The Blair government and the reshaping of social housing. *European Journal of Housing Policy*, **4**(2), 185–207.

Audit Commission: www.audit-commission.gov.uk
Housing Corporation: www.housingcorp.gov.uk
Information Gateway: http://www.sapling.org.uk/housing
National Housing Federation: http://www.housing.org.uk

11 A Toolkit for Engagement

11.1 INTRODUCTION

The preceding chapters have created a model for the engagement of the construction industry with clients and analysed the basis of this in detail. In addition, we have used the model to describe six client sectors, which provides critical introductory information about these sectors and demonstrates the approach of using the model. This chapter uses the model and approach to lay down a process of engagement with clients. By necessity, this involves some repetition of ideas introduced earlier and, to that extent, acts as a summary. We also comment on current approaches. None of these are wrong, and they can be used effectively for the better engagement of clients; however, they are not sufficient. They are an important part of your toolkit, especially as what we are proposing is quite radical and some form of transition is required into this approach. These current tools can be used as they are accepted by the industry; however, they can be adapted to develop the engagement towards a more successful outcome.

In creating this toolkit, we are conscious that individual clients are different from their sector norm, and each project involves different people. The basis of the toolkit is as a framework for action rather than a prescription for action. It gives an overall structure and reason for the approach, a sequence of activities, some details of these activities, and some remedial measures to action when the process is not working. Simply following the toolkit as a set of procedures is unlikely to deliver the results. What is required is an understanding of why the processes are necessary so that as events happen, someone can manage the engagement rather than the procedures.

Summary of issues

We start our summary by restating the basic thesis. Clients see and experience building differently from the industry. Building involves change in the client organisation. Buildings are large, and therefore the change can be large. Change puts pressure on the client organisation and its people as it involves actions and experiences outside the norm and exposes gaps and contradictions in these norms. In particular, the different objectives of the divisions within the client organisation induce conflict. This situation is made worse by the fact that building involves unknowns that are unformed, coming to existence only during the process of building, which limits the ability of plans to be delivered. The organisational change in the client and the experience of unknowns during building induces emotions in the client, which feed back to affect the situation in their process of building and in the organisational change. The industry is used to these changes and unknowns, but the client organisation is not.

In order to accommodate this, we have proposed seeing the client as a knowledge-processing system that has a known, rational formal side and an unknown, non-rational informal side. As a client comes to build, it exposes its values about building, about its organisation and about people. These values alongside the industry norms and values determine the means and ends of an engagement around change. The client change and building change are set within an external environment, which provides purpose for the change but also constrains it. Client satisfaction requires achievement in three areas: in the building, in the organisation and in the people. There are two change situations: the client organisation is changing to meet some organisational aspiration and the project is changing around the building. These interact and cause conflicts of purpose and of values. The process develops through the means rather than merely through the proposed ends set at the beginning. The ends and means involve both unknown and known components, and we have suggested that we consider these separately as two routes of change. These are, in fact, operating together, but the separation focuses attention on the unknown, which has not been considered in sufficient detail and thus not managed. The known route involves the conventional issues of building procurement and goal-directed change management in organisations, which are handled by formal management and procedures. The unknown route involves the conflicting objectives in the organisation, the conflicting objectives in building, and the conflicting objectives within people. Some of this is hidden or latent, which means that it could be determined if the process of enquiry makes it happen; some, however, is unformed, emerging only as the change transpires from the application of the means. The formal route and the informal route and the change in the project and change in the organisation interact and cause confusion. This enables unknown and compulsive manipulation of circumstances to occur, which are difficult to observe and difficult to control. This includes operational islands in the client pairing with operational islands in the industry around shared values, which can disturb the project and the organisational development by covertly directing them from the outside. Also, the emotional state of the client can induce a flipping between a power position, whereby the client is in directorial control, and a dependency position, in which the client is beholden to the actions of the industry. At its worst, this creates the adoption of persecutor and victim roles, which become part of the unknown means and can destroy the evolution of a successful project and organisational change.

As organisational and building changes take place, there are deviations between formal plans and reality and between aspirations and reality, regarding both the organisation and the building; this is shown in Figure 11.1. Thus, ultimately, four gaps in satisfaction are

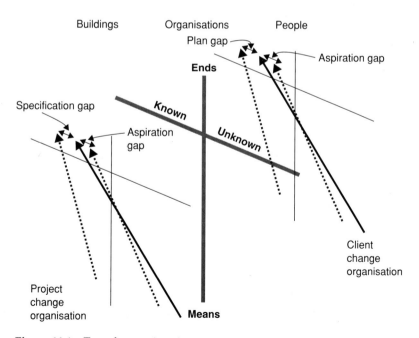

Figure 11.1   Two change situations.

produced – a plan gap and an aspirational gap in relation to the organisational end; and a specification gap and an aspirational gap in relation to the building end. These gaps create negative emotions because of the negative power of expectations, which means that people feel a loss of prediction and a loss of facility, which is difficult to overcome simply by the successful gains. Success is determined by the negotiation of the meaning of these gaps within the organisation and between the organisation and the industry.

Fundamentally, we believe that it is the responsibility of the industry to help the client. Later, we discuss clients that may not want to be helped. We need to develop abilities to understand clients and to manage the engagement between the industry and clients.

11.2 OUTLINE OF TOOLKIT

A toolkit needs to address the four areas that require to be managed, as shown in Figure 11.2. Each area has means and ends.

The normal approaches to construction handle the formal areas of the project change and methods of management in area 2. These will not be discussed here, but they can be accessed in other work (Barrett & Stanley, 1999; Kelly *et al.*, 2002; Winch, 2003). The formal and informal changes in the client areas 1 and 3 are the client's responsibility;

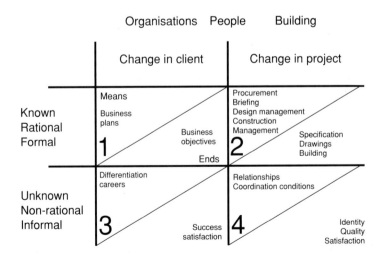

Figure 11.2 Areas to be managed in the client.

however, this may not be managed, particularly the informal route. The change and management (non-management) here are of great consequence to the project, both for the means and for the end success, and this becomes the industry's concern. Finally, the informal aspects of project change area 4 are the industry's duty; however, these are not managed by the formal system, and so an additional process is required.

This gives us three areas to be detailed for the toolkit:

❏ Understand the client's business (area 3).
❏ Work with the client's change processes (area 1).
❏ Manage the fragmentation in the industry (area 4).

Managing these will allow the current formal design and construction management processes (area 2) to be effective. The issue is to not let these formal processes dominate. As they are so well understood and so well ingrained in the industry's psyche, there is a tendency, when things get difficult, to return to this as the means of action. This is an anxiety-reduction, back-covering strategy; however, generally when people feel most inclined to use them is the time when it is best to avoid them. We deal with this in Section 11.5.

The most important and difficult part of the toolkit is working with the client's change processes. We deal with this by first using a process consultation approach. We have to understand this change process and help the client to manage it. This is not something that can be managed by diktat: the diktat would become part of the means, which would interfere with the unknown side and the known side in a destructive way. It is managed in the sense of coping and engaging with it, so that it might be possible to influence it, such that the implications for the project change are less unknown and there are relationships that can adapt to circumstances. The industry needs to develop skills in this area as well as undertake actions.

Understanding the client business involves understanding both the sector and the business environment and also the client's particular business knowledge and processes, including its history. Examples of this, for six client sectors, have been presented in this book. The process for doing this is outlined in Section 11.4.

The implications of the fragmentation in the industry need to be managed. Fragmentation has three consequences. First, it creates difficulty in the integration of the formal delivery of the building; second, it creates the potential for different objectives of the different industry organisations to disturb the project change; and third, it allows operational islands in the industry to be paired with operational islands in the client around their shared values, which disturbs the project and organisational development. It may be possible eventually to remove the fragmentation, but at this point the solution is to manage it; we discuss this in Section 11.5.

We will add a fourth area for the toolkit, which is concerned with developing the approach, which is discussed in Section 11.6.

This amalgam of processes that are required to enable a successful project and a satisfied client needs to be developed. This requires clients to learn as much as the industry; however, it is the skills of the industry as a change manager that are key, and new processes need to be created within industry organisations and skills developed in individuals.

11.3   WORKING WITH CLIENTS' CHANGE PROCESSES

We will use a process consultation approach to undertake this task. According to Schein (1987), 'Process consultation is a set of activities on the part of the consultant that help the client to perceive, understand, and act upon the process events that occur in the client's environment.' Process consultation is one of several approaches that management consultants use to help organisations to undertake change. It is part of an organisational development field (French & Bell, 1999), in which it is important that the role of consultant is to help others to help themselves. This is exactly the approach that is required in construction; indeed, given the power differential, it is the only form of engagement with a client. We use process in a dual sense: first, it involves an ongoing engagement to achieve results rather than a single action; second, it involves the social and psychological behaviour of groups. It is the latter sense that seeks to work with the unknown dimensions of organisations; however, the way of doing this is the ongoing engagement.

Process consultation was designed for management consultants; however, the industry's engagement with clients is not constituted in this manner. Process consultation assumes a single consultant and that the organisation has invited the consultant in to assist. As this is not the case with the industry, we reconstitute process consultation as an informal role of the industry. Ultimately, we believe that this could develop as a key industry skill, for which it would get recognised and paid (Dawson, 2000). We also believe that many construction project managers may adopt this role, even though their contract is about task management. There are two important aspects of the reconstitution of process consultation for the management of the engagement with the construction client: first, it is about the form of industry access to the client; second, it is about ways of working with the client. Access has to be beyond the contractual divide, otherwise the client remains in the unknown of the industry and cannot be helped. In addition, access needs to be available to not only one part of the construction industry but to a wider involvement, otherwise the learning is held too closely in a single relationship. This wider access requires acceptance on the client's part and acceptance from within the industry. The sequence of engagement is shown in Figure 11.3, where it refers to other activities associated with the project. We now discuss in detail the method of this engagement. The sequence has six phases of engagement, although phases 3, 4 and 5 may be repeated. This diagram also shows all three areas of change processes that were identified in Section 4.7, namely the client organisation, the project and the industry. Although we have not discussed this, it is important to understand that the industry changes with each project and with the ongoing industry development. This industry change may need to be managed as well as the other locations of change, but it can be part of this process consultation approach.

Although there are six phases during this process consultation engagement, the most important point is at the beginning. Neumann (1994) discusses the difficulties of the beginning of a client–consultant relationship. She identifies the three early phases as scouting, entry and contract, and diagnosis and strategy for engagement, which we include as our first three phases. Scouting is early engagement, when the consultant checks the territory and identifies dangers. At this stage, the client has already started to change, as the aspirational idea has surfaced and been partly disseminated. The objective is to enable a return for another meeting with as little ground disturbed as possible, but with information acquired through observation. In these discussions, it can be ascertained as to whether the process consultation can be an explicit joint enquiry or whether it has to be hidden as the client is not interested in such help. If the client is too difficult, then there is always the option of walking away from the contract, but this is not the

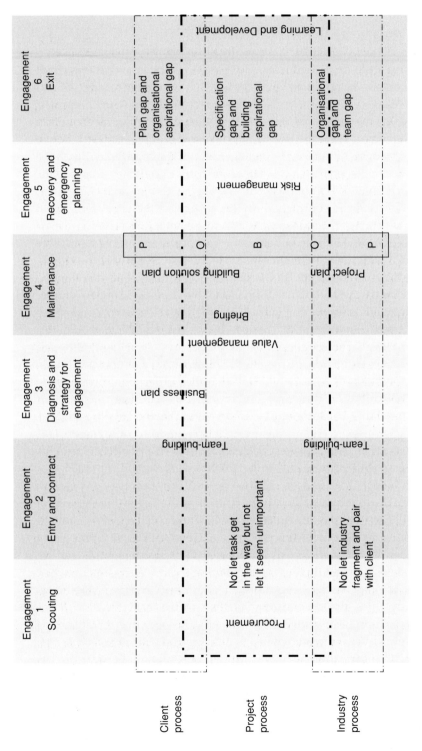

B, building; O, organisation; P, people.

Figure 11.3 Sequence of process engagement with client through project.

normal approach of the industry. The entry and contract phase refers to an official engagement to present the ideas of process consultation. How this appointment is created is important, as is the degree of difficulty of obtaining it. During the meeting, the needs for process consultation and the approach will be elicited, in particular the client's view of the situation, as to whether it is seeking an expert to act on its behalf, a pair of hands to act under it, collaboration to act with it, or mutual engagement when the client is seeking to enter into a joint learning process (Neumann, 1994). When buy-in and agreement are reached, a social contract is formed; the basis of this is moral rather than legal. The diagnosis and strategy for the engagement phase takes the organisational format and the project format and determines the process difficulties (but not task difficulties), so that the start of an ongoing dialogue of enquiry can continue. Although the most important information is received and the relationship set up at the beginning, the engagement does need maintenance. This involves a regular meeting depending on the rate of change, which, of course, can vary throughout a project. In addition, it is useful to discuss with the client an engagement recovery approach in the format of an emergency plan. This cannot be done when things have gone wrong, as then emotions are too high; thus, a more abstract plan is created in order to allow continuing dialogue. Finally, an exit approach needs to be discussed and planned; indeed, it is through this exit approach that the gaps in plans, specifications and aspirations can be managed so that there is project success and a satisfied client.

In Figure 11.3 we also place current industry techniques, which we believe align with this approach. There needs to be a positive approach to procurement, in which the client sees its role in the approach to building as being important, not as separate worlds. In phase 2, the client has to accept the help of such an approach. Then team-building of the industry participants needs to take place in order to prepare the industry to be able to help the client and to work across its fragments. In this way, the industry can then proceed with a team-building session with the extended client. The client's entire business plan does not need to be revealed in phase 3, but the objectives and assumptions behind the change processes need to be understood. The processes of value management (Green, 1994) are particularly important when a building solution is being proposed and the details of the values behind this need to be appreciated. This should occur before briefing about the building, as this should be based on the particular need for a building to undertake the organisational change. It may emerge that the client does not need a building solution and the industry needs to have the courage to reveal this, regardless of its compulsions to continue with a building. Both value management and briefing can start to establish the end values, which will evolve to be the criteria against which success is measured. The project plan and business solution plan acknowledge that change is taking place in two situations. In the maintenance phase, we are variously managing our three achievement areas – people, organisation and building – referred to as P, O and B in Figure 11.3. We have indicated P and O on the industry side to make us aware that the objectives of the industry need to be managed as well, otherwise the organisational fragments will be pulled apart. Recovery and emergency planning are not a normal approach of the industry (Boyd and Weaver, 1994) but can be regarded as a form of risk management; however, they are concerned with managing the breakdown in communications.

Process engagement

In a process consultation, the key understanding is that the means are more important than the ends. Our assumption is that if we manage the process (the means) of engagement successfully, then the product (the ends) will be successful. It is slightly contradictory with goal-directed change in the formal route, which assumes that getting the goals correct will enable them to be achieved. In addition, the methods of the industry are set around products (ends), and therefore it is difficult to move to a process approach without upsetting the system. In fact, architecture has often worked like this, in that the client does not buy an end design but the processes of engaging with the architect in design. Also, new approaches to briefing accept a process approach, whereby the brief is not complete at the beginning but is refined through a process of engagement during a project (Barrett and Stanley, 1999). As soon as we appreciate this, we see that it is the means of engagement and the means of engaging that become critical. We still have to deliver the product, but that added value, whereby the product meets the needs of the client, is determined by the engagement. Through this, the industry can align the development of the detail of the client's aspiration with the detail of the building, such that they are not too dissimilar. These, to some extent, develop together in a process approach, and therefore there is great opportunity to align them, i.e.

both can change so that they can converge. This is significant because, as servants to the client, the industry is not only doing what the client tells it but is developing the client; thus, the industry has a greater responsibility than simply the building, as it is also developing the client's aspirations. Thus, the product in the client's eyes at the beginning can be, will be and needs to be different from what the client gets at the end in order to obtain success and satisfaction. Clients with a belief in or an institutional need for certainty, like the public sector, may not like this and this problem will require to be managed.

Skills for operating a process approach to engagement need to be learned. The problems identified in this area involve both access to information and methods of management. Our objective is to help the client through its change process, so that its internal operations are effective and its engagement with the project is positive. The approach requires the industry to enquire into these aspects of the client with a view to gaining information about the state of the issues so that the industry can modify its action accordingly, and draw the client's attention to these aspects and their state so that the client manages them. By enquiry, we mean observation, asking questions and listening, and then feeding back further questions in order to delve into subjects. While this is happening, the client has to bring its attention on to the issues, thus making the client aware of them and, with luck, learning from them. This approach can be referred to as 'qualitative research' (Mishler, 1991), 'knowledge elicitation' (Sparrow 1998), 'process consultation' (Schein, 1987), 'catalytic intervention' (Heron, 1990), 'cooperative inquiry' (Heron, 1996), 'dialogue' (Isaacs, 1993), 'situational interviewing' (Latham *et al.*, 1980), 'appreciative enquiry' (Whitney & Trostem-Bloom, 2003) and 'action research' (Reason & Bradbury, 2000). Questions need to be phrased such that they are not evaluative of the situation and do not display emotion about a direction of the answer. It is possible to enter the enquiry using an example from another organisation and ask whether this is happening in the client's organisation.

In particular, the process consultation technique is similar to Heron's (1990) catalytic intervention with a client, whereby the aim is to elicit self-discovery, learning and problem-solving. Heron describes various states of how clients choose to act, including deranged, compulsive, conventional, creative, self-creating and self-transfiguring. Of particular interest are the compulsive, conventional and creative states, which, respectively, refer to rigid, maladaptive and repetitive

Table 11.1 Method for process consultation.

Negotiate a time commitment and stick to it
Ask open questions
Give the client as many opportunities as possible to clarify issues
Avoid debating matters; respond by saying 'Tell me more'
Use prompts such as:
   'Tell me about when x happened'
   'What do you remember most?'
   'What was it like to do x?'
   'What do you think happened?'
   'What was your experience?'
   'What was good about the situation?'
   'What made you feel uncomfortable, and how did you deal with that?'
   'Did anything surprise, worry or please you?'
Try not to fill any gaps by speaking; allow silences and let the client talk

behaviour; unreflective and conforming behaviour; and autonomous, committed and innovative behaviour. Catalytic intervention works effectively with the creative state and the conventional, where a degree of cooperative engagement is possible. The technique involves questioning and listening, and the process consultant works with the words and emotions of the client. An outline of the questioning approach is given in Table 11.1. Helpful actions include echoing words and phrases to check accuracy, paraphrasing to check meaning, and empathetic divining to solicit the feelings behind a statement.

The process approach shown in Figure 11.3 needs to be created with the client rather than being selected. The following areas need to be enquired into using the methods outlined in Table 11.1, and managed throughout the engagement:

❏ engagement
❏ pressures from the external environment
❏ gaps and contradictions
❏ unformed outcomes
❏ individual's emotions
❏ end gaps.

Together these constitute the management of change. The client needs to see building as a process of organisational change by being asked questions about the effect of the building on its organisation. A set of questions in the above areas is shown in Table 11.2.

Table 11.2 Question areas for enquiring into the client's change process.

Question area	Objectives	Questions
Engagement	Elicits client's feelings about the way the industry is working with it	What is your experience of past building projects? What are the problems of change in your organisation? How will the building assist your organisation to develop? It is our experience that organisations have problems when they build; do you see this as a problem in your organisation? If we see your organisation having problems, should we tell you?
Pressures from external environment	Elicits effects of the different external organisations and how these are managed, e.g. market, competitors, suppliers, regulation, politics, finance	What are the drivers of your market? Who do you compete with? What are their key competitive points? Who are your suppliers? What problems do you have with your suppliers? What regulations apply to your business? How do they affect it? What politics surround your environment? How do businesses in your sector fund development? What are these funders' concerns? Who is responsible for observing the external world? How are problems managed?
Gaps and contradictions	Elicits degree of splitting of the organisational divisions and how these are managed	What are the structural and operational divisions in the organisation? What are the differences in the objectives and ways of operating in these divisions?

Table 11.2 (*Continued*)

Question area	Objectives	Questions
		How do the procedures overcome the divisional boundaries? How are decisions made?
		How are problems between divisions managed?
		Where is strategy devised, and how is it communicated and implemented?
		How is strategy resolved against what it is possible to do?
		How are organisational and divisional identity maintained against change?
Unformed outcomes	Elicits attitude to uncertainty, whether such outcomes are emerging, and how these are managed	How many processes are designed and time spent to ensure that outcomes are goal-driven?
		How does your organisation view it when outcomes differ from goals?
		How does your organisation view errors, mistakes and deviations?
		How are procedural improvements surfaced, accepted and implemented?
		How do your organisational procedures deal with changing environment?
		What management actions take place when deviations from plans occur?
		How are informal methods of resolving problems viewed?
		Do different divisions view unpredicted outcomes differently?

Individual's emotions	Elicits aspirational drives in the organisation, results of the change, what drives its aspirations (organisations, people), what drives it apart, what annoys it	What gets rewarded in your organisation? What would be regarded as a good improvement?
		What are the symbols of success and achievement?
		How does the organisation view individuals' achievements?
		How do gaps between people's careers, beliefs and desires and the organisation get surfaced?
		How do people get promotion? How often does this occur?
		How do people compete to get promotion?
		What issues cause people to disagree? How do people behave when they disagree?
		How are people different in the different divisions?
		How do people in different divisions view external people working for the organisation?
End gaps	Elicits viewing of the deviations from the plan and aspirations that are emerging; it is the way of managing the meaning of them so that the client is satisfied	What would be regarded as successful outcomes for the business?
		What would be regarded as problematic outcomes?
		What would be regarded as a satisfying process? What would be regarded as a non-satisfying process?
		How does the overall success get interpreted within the organisation?
		Which divisions are primary in the interpretation of success?
		If individuals are unhappy about the process of change, how do they express it in the organisation?
		What processes does the organisation use to rationalise non-success?

Operations of process engagement

The questions presented in Table 11.2 fit into the following context. As we stated, the objective of a process engagement is to help the client cope with its change process so that its internal operations are effective and its engagement with the project is positive. Different divisions in the client organisation will see the change differently and will be part of the problems within the client system. These differences may be hierarchical or functional, as discussed in Chapter 3, but Schein (1987) identifies six other categories associated with the interaction with the change. The *contact client* is the first contact and often the site of the immediate relationship. This may be a director or someone from the facilities department or even another consultant. The *intermediate client* comprises those people who get involved in the decision-making because the project impacts on them, because it expands their job (e.g. the IT department needs to have cables) or they use the new facility as a service (e.g. an X-ray department) or they administer it (e.g. finance). The *primary client* is the division that ultimately owns the problem and that is getting the extra facility to expand its operations or take over another outlet. Primary clients are different from intermediate clients because they have a direct stake in the success of the project. The *unwitting clients* are members of the organisation who will be affected but are not aware of the implications or are not consulted. The *ultimate client* is the whole organisation that embodies the strategic impact that feeds around the organisation. The *involved non-clients* are those tangentially involved but surround any project, such as neighbours. They will not have a contract, but they can often interfere through administrative or regulatory consultation processes but cannot be ordered around. The approach to mapping the client environment that we used in the sector analysis, and which we discuss in the next section, can be modified easily to identify these different clients. An awareness of these different clients allows their important differences to be managed.

The degree of change induces different emotions depending on the normal view of uncertainty by the client, and this can be observed during the process consultation and, if at all possible, not reacted to in a negative way. Neumann (1994) relates how organisations create social defence mechanisms in order to avoid experiences of anxiety, doubt and guilt. This induces what Bion (1961) referred to as basic assumption behaviour. This involves three responses: fight/flight, dependency and pairing. Fight/flight from the group involves fighting or fleeing from something; the 'something' may be the construction industry, but the important thing is that the overreaction preserves the group. Dependency from the group occurs when it seeks a leader to act and provide for it; again this can be the industry, and the danger is that the group starts to be totally reliant and unable to act itself, and has high expectations of the leader. The pairing response involves the group or individuals seeking a similar partner in order to re-create the group; this most often allows avoidance.

The state of clients can be defined as underbounded, overbounded and balanced, as introduced in Section 4.6. This comes from the ideas of open system theory and relates to the degree of interaction between an organisation and its environment across its boundary (Alderfer, 1979). Overbounded organisations do not respond to changes in their environment. They do not develop to address external changes and are unaware of external needs. Underbounded organisations respond instantly to changes in their environment in such a way that they lose purpose and any effectiveness of action. Underboundedness causes meaninglessness and value conflict over objectives, wasteful utilisation of resources, fragmented authority relationships, diversity of expectations over roles, problematic communications due to withdrawal from relationships, and a propensity to anger and conflict. Because of this, intergroup dynamics become concerned with identities, and there is withdrawal into role in the face of conflict, reflecting fight/flight behaviour.

The underbounded organisation will be affected by everything that is going on in the project and in its business environment. This means that it will change its minds regularly and find it difficult to make internal decisions because of fight/flight responses. The organisation will be used to change but have no mechanism to manage it effectively (Alderfer, 1979). The industry may be seen as an ally, but this can be a volatile relationship, depending on the environmental pressures and can induce power and dependency flipping as suggested in Section 4.7.

Overbounded organisations are not aware of what is happening in the project, being content to work on the initial goal prescription; nor do they have much awareness of their business environment (Alderfer, 1979). They are slow at responding to the industry, expecting bureaucratic processes to prevail. They are heavily disturbed if anything significant emerges that was not predicted and will search out a leader to be dependent on. The industry should avoid taking this role compulsively without establishing the consequences with the client.

Overbounded organisations are reluctant to talk about their organisations and are unlikely to enter openly into process consultation.

It is normally assumed that only small, inexperienced clients are compulsive and that large clients are well managed and have secure processes. As we have shown, there are gaps and contradictions in all clients. Larger clients, however, tend to be overbounded and have more gaps and contradictions, as there is more scope for these. In smaller clients, these are more concentrated with individuals and closer to the organisational purpose. In large clients, the corporate nature creates differentiation and alternative purposes, which make decisions difficult and drive away reality from aspirations. The processes of procurement can encourage these compulsions. Overbounded clients believe that they are protected by contract and find it difficult to make decisions that cannot be known with certainty. Boyd and Wild (1999) argue that overbounded organisations become underbounded as a result of building because their systems cannot handle the unformed unknowns. Overbounded clients do not like change and have a dependency on the management of systems, which are concerned with stability. They will not have the conceptual or emotional tools necessary to deal with change. In addition, their expectations of certainty come from their bureaucratic management experience, and they may believe that a construction project should be the same. Thus, there is a two-fold problem with overbounded clients: no tools to deal with change, and expectations of activity only from a bureaucratic perspective.

The type, extent and depth of change and how it is managed in the client are important and are what we need to find out about. The more fundamental it is, and the more it requires or involves a change in conception, the more difficult it is and the more likely that there will be problems. It is in these projects, and at particular times in them, that uncertainties are highest, emotions are highest and most sensitive (non-linear), conflict is endemic and issues are unexpressed. Alderfer (1979) suggested that different forms of intervention are required for an over- or underbounded organisation. This can be determined at the diagnosis stage of the process consultancy. Underbounded organisations require an approach that gets them to create structure and to agree on key purposes. Overbounded organisations require an approach that motivates them to look outside their boundaries and that legitimises emotions.

The overall process consultation side of the management of the change in the client and the building change can be regarded as evolution management (Van der Erve, 1997). Van der Erve refers to such people as our process consultants, who can facilitate change, as creation catalysts. These consultants have a view of the whole but balance the autonomous actions of the client by ensuring the client's awareness of the fact that success requires its working with the interdependence of the wider project and business organisation. The process consultant can also help with what Van der Erve refers to as the reincarnation of the organisation, which means that the change is seen positively and is part of a new beginning for the organisation. In this way, the outcomes, whatever the plan, specification and aspiration gaps, are seen in a positive developmental way and as part of a successful future.

Working with people

As the industry works with a client, it will meet many different people from the client's business. These differences are important, as outlined in Section 3.5. The industry needs to understand these differences of role, function and personality, in order to work positively with them. These differences not only determine how the various individuals perceive the problems but also can indicate how the individuals will react emotionally to changes as they occur. The changes can be both an opportunity and a problem for individuals, who can then influence the project or react to the project in unknown and apparently non-rational ways. The process consultation needs to acknowledge and deal with individual differences by making them explicit and exploring the consequences of them. This should come out of the question inquiry. A balance needs to be struck between not avoiding differences and not searching them out obsessively. When sensing-judging people are working in an overbounded organisation then issues of blame, distrust and finding fault can make the project chemistry (Nicolini, 2002) non conducive to positive engagement. Conversely, when intuitive-perceptive people are working in an underbounded organisation then unclear goals, shifting meaning and dependency behaviour can disturb progress. Organisations which do not, or cannot, acknowledge individuality tend to hide differences which will become exposed when anxieties and unformed unknowns occur. Understanding and expecting this can help to avoid conflict and work towards a positive solution to issues rather than getting caught in a personal battle. It also can help to anticipate how individuals' expectations will be satisfied by the outcomes of the project.

11.4 UNDERSTANDING THE CLIENT'S BUSINESS

Understanding the client's business involves determining the known aspects of the business environment, the purpose or service of the organisation, the structure of the organisation, and the defining processes of the organisation. This requires an enquiry through published and online resources in order to create the sort of understanding displayed in the six sector chapters of this book. These are known and formal areas; however, aspects of gaps and contradictions in operation can be inferred. Thus, it provides useful information for the process consultations and probably will be undertaken between the scouting and entry and contact phases. The objective of this is to determine the client's knowledge and how and why the client would respond to different situations. A diagram to assist this enquiry for a private-sector client is given in Figure 11.4 and for a public-sector client in Figure 11.5. If the client is amenable, as determined from the process engagement, then this can be done with the client. In our interviews, we used these diagrams to get the clients to discuss their businesses; this causes

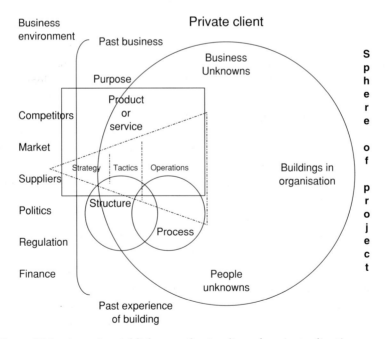

Figure 11.4 Areas to establish an understanding of a private client's business.

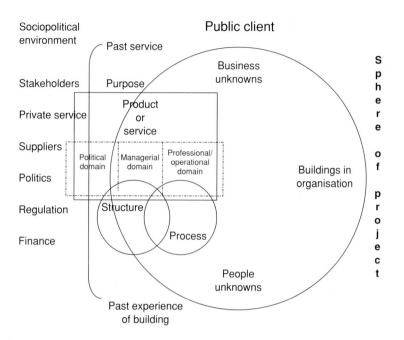

Figure 11.5   Areas to establish an understanding of a public client's business.

the client to reflect on an holistic view of its business rather than the functional view that most often individuals' roles require.

The diagrams have three areas: the external world (i.e. the business environment or the sociopolitical environment), the constitution of the organisation (i.e. the purpose, structure and processes), and the sphere of influence of the building project, which interacts with the purpose, structure and processes of the organisation and also the organisation's people. The organisational differentiation, either business hierarchy or the public service domains, are superimposed on the constitution of the organisation. The first two were described in detail in Chapter 3 and the third in Chapter 4.

The influences of the external world were discussed in Section 3.3. Both public and private clients are influenced by suppliers, politics, regulation and finance. The private client has to contend with competitors and markets, whereas the public client has to contend with stakeholders and private providers. Both have suppliers that are part of a value chain and thus dependent on the client but are also in competition with the client for higher prices. Both are constrained by regulation, which covers a number of areas, including business law, employment law, health and safety, building regulations, environmental legislation and development control planning. Although both have political pressures, these are of a much higher magnitude for the public client and place major constraints and disturbances on any single public-sector service deliverer. In a similar way, both are influenced by finance; in the private sector, however, this refers to equity or debt funding and the wider national financial context such as interest rates and taxation, but in the public sector it refers to treasury and European Union (EU) arrangements for capital funding and public–private financing arrangements. Private businesses operate within a market for their goods or services and against competitors. Public clients provide services for sundry stakeholders, often having to decide who to disadvantage in favour of another and thus inducing many external pressures from conflicting groups. Although public clients do not have competitors, new arrangements for the delivery of public services mean that there are pressures to provide these from private suppliers.

The constitution of the organisation involves not only structure and processes but also how it is viewed as a legal entity in the wider business world. This constitution constrains it but gives it a format to undertake its business. The purpose of the organisation needs to be explored as this determines its values and strategy (see Sections 3.3 and 4.2). It is useful to consider some history of the organisation, as past events can leave a trace in current actions from past successes and failures that may not be understandable otherwise. The structure and processes determine how the client goes about its business and who undertakes specific roles. In both diagrams, we have acknowledged vertical differentiation, in the form of strategic, tactical and operational for the private client, and in the form of political, managerial and professional for the public client. This is important as it provides us with some idea of the gaps and contradictions that are there or that will be aggravated when building occurs. Indeed, it is useful to explore the processes of communication across this vertical differentiation and across the functional differentiation and to enquire about communications and decision-making problems across organisational islands.

The external influences and the internal differentiation can be usefully mapped in a diagram form, as shown in Figure 11.6. This was used to present information on the client sectors, but the compilation of them allows a degree of analysis of the operation of the organisation and, if constructed with the client, can assist in appreciating the world

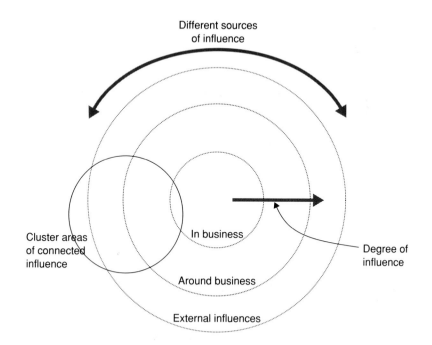

Different sources
of influence

Cluster areas
of connected
influence

In business

Degree of
influence

Around business

External influences

Figure 11.6   Format for mapping the influences on the client's business operation.

as seen by the client (see Section 3.2). The diagram has at its centre the core business; differentiated sections, stakeholders and other agents are placed at varying distances from the centre, depending on their degree of influence on the business. Those close in are most influential within the business, those on the next annulus tend to be immediately interacting with the business, and the outer annulus has the external world. It is useful to cluster entries that operate together, as this allows a visual representation of effective influence and interaction.

We need to determine the sphere of influence of the potential building project. This identifies where the organisational change will be taking place and thus allows us to distinguish the different types of internal clients. These may be declared in a business plan and an appropriate management strategy can be implemented; however, it is important to establish this and to enquire about its assumptions. The map described above can be used to present this by using a line of a different colour. It as at this point that we start to see the business unknowns surrounding the building project and the organisational change project. In addition, the organisation's attitude to building can be established, particularly

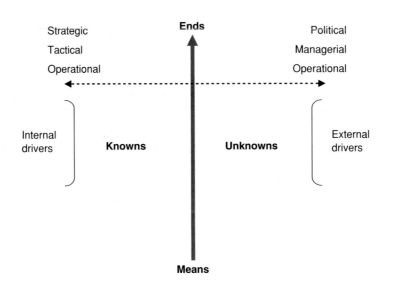

Figure 11.7   Means-and-ends diagram of client drivers and values.

if it has had some bad experiences in the past. It is useful to collect names of people within the sphere of influence and to appreciate their stake in the organisational and building change. Again, this starts to interact with the process consultation, and much of this can be done simultaneously in suitably accommodating organisations.

Finally, we need to draw together the drivers and values of the client's business and determine the client's approach to change. This can be usefully presented in a means-and-ends map, as shown in Figure 11.7. This was used in the sector chapters in this book.

The known side starts with the internal drivers including the organisational goals and processes, however, these also contain unknown and contradictory aspects. The unknown side starts from the external drivers which interact with the means which cope with these pressures, some of which are known. It is useful to present issues as dyads in the sense that an issue on the known side has a corresponding issue on the unknown side. In addition, the certainty of an issue can be represented by the position on the spectrum between known and unknown. It is useful if this is done with the client, as this helps to establish the client's understanding and emotional significance of the issues. The ends at the top can be differentiated between strategic, tactical and operational for the private sector, and between political, managerial and professional/operational for the public sector.

11.5   MANAGING THE INDUSTRY'S FRAGMENTATION

The fragmentation in the construction industry is legendary and is blamed for its failures in the delivery of building and in satisfying the client (Emmitt & Gorse, 2003; Groak, 1992; Walker, 2002; Winch, 2003). Fragmentation means the participation of different organisations in order to undertake the large and complex task of building. The fragmentation is both in time, where different organisations undertake different parts of the task as it develops, and in function, where many different organisations are involved at any one time. Thus, not only are clients differentiated, as we discussed in Section 3.3, but also the industry is differentiated; in a similar way, this causes gaps and contradictions that prevent adequate integration (Emmitt & Gorse, 2003). The situation is worse in construction, as the operational islands that are formed have a clear identity and are constituted as professions or trades. Thus, they can have more allegiance to their profession or trade than to the project and can operate in projects with very individualistic objectives. The operation of this complex supply coalition generates behaviours that are difficult to understand and difficult to manage (Winch, 2003).

Among all the other problems that the client has when it builds, this is an additional problem that the client does not want to have to cope with; nor does the client want to pay for the consequences. Many of the current initiatives to improve construction are designed to overcome this problem. These initiatives split into removing the problem and managing the problem. Design and build, prime contracting and offsite manufacture are attempts to remove it. Supply-chain management, process-mapping, project management, partnering and team-building are designed to manage it. It is probably the case that completely eliminating the problem is not possible, as there are both strong economic drivers and strong social drivers for its existence. Therefore, an approach to managing it needs to be utilised.

Fragmentation creates three problems: first, a difficulty in integrating the tasks required to deliver the building; second, a conflict of organisational objectives between the operational islands and the project, such that those of the project are secondary; and third, that operational islands in the industry get paired with operational islands in the client around their shared values, which disturbs the project and organisational development. The last of these comes out of our analysis presented in Section 4.7 and contributes to further fragmentation of the industry team. These can be regarded as process problems in a similar way to the client's change process problems. Thus, they require to be, and can be, managed by a form of process consultation to the industry coalition. In fact, we included this in our process consultation phase diagram in Figure 11.3. Remember that there are compulsions on both sides that need to be surfaced through a process of enquiry.

It is tempting to leave the industry side to muddle through (Riddick, 1992) as it has always done and to concentrate on the process consultation in the client. However, the situation of the engagement is a lot more interdependent than is illustrated in Figure 11.3. Thus, problems emerging from the industry's change processes affect the project processes and the client processes. It is worthwhile considering all of these change situations as underbounded systems, as we described in Section 11.3. Thus, problems permeate from one system to another, and the systems react strongly. The reactions to underboundedness induce fight/flight behaviour, and this ricochets around the industry organisations, the project and the client organisation, causing impacting of information and hardening of positions, i.e. people start working on expectations and perceptions rather than on real information, objectives are individualised, and blame and protection become the norm. This mistakenly provokes simple explanations of cause and effect and the allocation of blame, which make matters worse. This dynamic is latent and is excitable by issues occurring anywhere or indeed by worries of issues occurring anywhere. This needs to be managed in each system but particularly in the industry, which should be operating to strengthen the client in its change.

11.6   DEVELOPING THE APPROACH

With any new approach, as has been outlined in this book, the processes of operation need to be learned and skills developed. This book provides only an understanding of the issues and a framework for action. The detail has to be extrapolated from this for the particular circumstances found. The experience from using the approach also generates new methods and an awareness of processes that do not work. This is critical for the development of the approach and so is part of the toolkit. We recommend that your organisation keeps a log of observations and also conducts a review at the end of the project. In addition, the success of the approach stimulates the industry to keep trying, and therefore it is important to document and broadcast the outcomes. If you start achieving, celebrate this success: it reinforces the approach. The approach can then be part of a training course and others can work with it.

The process consulting approach benefits from working closely with the client. This learning *about the approach*, in order to improve it, should be conducted with the client. The client's perspective and insight are crucial, as it is their experience that we are seeking to satisfy. Their general observations and their presence at the final review will give useful feedback and help to create a longer-term relationship that might be useful in future business. However, there is an additional benefit of the approach: learning *from the approach*. The nature of using an enquiry method into complex industry or client systems is that these are analysed and their effectiveness established. Thus, the approach helps to educate the client about itself, so that both the industry and the client can learn about developing their businesses. This could be as additional service offered by the industry; a strategy suggested by Dawson (2000).

The skills of process consultation are not immediately available from within the industry. As we have said, we believe that some project managers undertake something similar, although this is hidden by their task-integrating function. It is possible for any discipline to undertake it; however, it needs a degree of separation from one's task and also needs access to a wider client and project organisation. In a more mature industry and client setting, this could be discussed and the role allocated. Certainly if an individual takes on the role they should try to explain this to the industry team and the client. It requires a strong, confident and socially capable individual who can see wider issues and is willing to ask difficult questions. It could be an external consultant; however, the economics of this are unlikely to be acceptable to a client.

One functional department within construction industry organisations that might be able to take on this task is business development. Many contracting and large consulting practices have business development managers who watch their business environment, seek business and market the firm. They tend to be socially capable in addition to having industry skills. This process consulting role can be seen as part of long-term marketing and client relation management, which falls within their remit. They are also slightly distant from the project itself, being in the wider organisation, and can look more dispassionately at events. Their role – to seek business from clients – is akin to the scouting phase of the process consultation approach, and this provides them entry at a number of levels. Having some continuity of their engagement can only improve the relationship with a client or handle any problems that emerge from the project.

It is also necessary to learn to deal with the client that is unwilling to engage in process consultation or unwilling to be helped. This is a warning sign about problems, certainly for the project but possibly also about the management of the change in the client. Such clients are overbounded, believing that their world is secure, that the external world has nothing to tell them, and that they are in command of the actions of the world. It is possible to work with these clients in the old way of requiring all information up front, not accepting positive emergence, and protecting oneself with contract. It is possible to make money when their overbounded state meets reality; however, this is reproducing bad practice in yourself and in the client. Heron (1992) refers to these as compulsive states. He suggests that a confronting intervention may be effective in these circumstances. This is difficult, as it involves a challenge but requires a language of support. It is easier if some factual aspects of the failure to engage can be related so that reality cannot be avoided. It is important not to get into an argument, and the simple reiteration of the benefits is better than defending or attacking particular issues. The process consultant is taking power, but if the intervention degenerates this can lead to the client taking a persecutory position (see Section 4.6).

Clients are being entreated to join the Clients' Charter, which involves clients taking leadership, working in integrated teams, seeking whole-life quality and respecting people (CCG, 2005). This, we believe, is a good thing. Such clients will be primed to accept process consultation through their commitment to teams and people. Indeed, their leadership in these matters may require the use of process consultation in order to ensure an effective engagement. We hope that negative clients can be identified and that these will find it more difficult to employ the best of the industry. Negative clients need to pay for their maladaptive processes, which constrain effective development. We also believe that as the technique is successful, clients will accept it and will reward it. In such a way, this will encourage others within the industry to learn to use the approach.

11.7   CONCLUSION

Our difficulty in understanding clients is two-fold. First, the area is multidimensional, dynamic and contradictory. Second, there are current conceptions and approaches that provide a different understanding. Thus, as you look at what this book suggests, you will most likely see it from the current position. Our task was two-fold: to communicate a new model and to persuade you to use a new model. The difficulties (multidimensional, dynamic and contradictory) give us some problems. The fact that it is multidimensional means that it is complicated and requires different ways of approaching different areas. The need to make it comprehensive and to deal with all clients in all situations required us to be rather abstract at times, which can appear distant from real problems. The fact that the industry–client interaction is dynamic means that, to some extent, it is changing, and so looking back things were different and looking forward things will be different. You can always find examples that do not quite fit the model, or at least change the significance of the parts. Again, to accommodate this we have had to be somewhat abstract in our representation. Finally, the fact that it is contradictory to previous and current conceptions makes the approach look wrong immediately. This gets us back to the problem of persuasion. If the industry and clients are unwilling to try it, then we and they will never know whether it works. What is required is a demonstration of the approach in order for it to develop and in order for it to be seen to be practicable.

REFERENCES

Alderfer, C.P. (1979) Consulting to underbounded systems. In *Advances in Experimental Social Process*, Vol. 2, eds C.P. Alderfer and C. Cooper. New York: John Wiley & Sons.

Barrett, P. and Stanley, C. (1999) *Better Construction Briefing*. Oxford: Blackwell.

Bion, W.R. (1961) *Experiences in Groups and Other Papers*. London: Tavistock Institute.

Boyd, D. and Weaver, P. (1994) Improving the management and operation of refurbishment projects. In *Proceedings 10th Annual ARCOM Conference*, eds R.M. Skitmore and M Betts. Salford: University of Salford, pp. 54–63.

Boyd, D. and Wild, A. (1999) Construction projects as organisation development. In *Proceedings of the 15th Annual Conference of ARCOM*, ed. W. Hughes. Liverpool John Moores University, pp. 221–229.

CCG (2005) The Clients' Charter. Abingdon: Construction Clients' Group. www.clientsuccess.org.uk. Accessed 28 March 2006.

Dawson, R. (2000) *Developing Knowledge-based Client Relationships*. Boston, MA: Butterworth-Heinemann.

Emmitt, S. and Gorse, C. (2003) *Construction Communication*. Oxford: Blackwell.

French, W.L. and Bell, C.H. (1999) *Organisation Development*, 6th edn. Englewood Cliffs, NJ: Prentice-Hall.

Green, S.D. (1994) Beyond value engineering: SMART value management for building projects. *International Journal of Project Management*, **12**(1), 49–56.

Groak, S. (1992) *The Idea of Building*. London: Spon.

Heron, J. (1990) *Helping the Client*. London: Sage.

Heron, J. (1996) *Co-operative Inquiry: Research into the Human Condition*. London: Sage.

Kelly, J., Morledge, R. and Wilkinson, S. (2002) *Best Value in Construction*. Oxford: Blackwell.

Isaacs, W.N. (1993) Dialogue, collective thinking and organisational learning. *Organisational Dynamics*, **Autumn**, 24–39.

Latham, G.P., Saari, L.M., Pursell, E.D. and Campion, M.A. (1980) The situational interview. *Journal of Applied Psychology*, **65**(4), 422–427.

Mishler, G. (1991) *Research Interviewing. Context and Narrative*. Cambridge, MA: Harvard University Press.

Neumann, J. (1994) Difficult beginnings: confrontation between client and consultant. Paper presented at the International Consulting Conference: What Makes Consultancy Work?, South Bank University, London, 29 January 1994.

Nicolini, D. (2002) In search of 'project chemistry'. *Construction Management and Economics*, **20**(2), 167–177.

Reason, P. and Bradbury, H. (eds) (2000) *Handbook of Action Research: Participative Inquiry and Practice*. London: Sage.

Riddick, J. (1992) The cult of muddling through: application of stratified systems theory in understanding construction project organisations. In *Festschrift for Elliot Jacques*, eds S. Cang and K. Cason. Arlington, VA: Cason Hall.

Schein, E.H. (1987) *Process Consultation*, Vol. II. Reading, MA: Addison-Wesley.

Sparrow, J. (1998) *Knowledge in Organizations: Access to Thinking at Work*. London: Sage.

Van der Erve, M. (1997) Evolution management. In *Strategic Change*, ed. C. Carnall. Oxford: Butterworth-Heinemann.

Walker, A. (2002) *Project Management in Construction*, 4th edn. Oxford: Blackwell.

Whitney, D. and Trosten-Bloom, A. (2003) *Power of Appreciative Inquiry: A Practical Guide to Positive Change*. San Francisco, CA: Berrett-Koehler.

Winch, G.M. (2003) *Managing Construction Projects*. Oxford: Blackwell.

12 Postscript

This is a complicated book. It is something of a work in progress. The research for this book, like all projects of change, can be understood by our means-and-ends model involving knowns and unknowns. There is a set of ends that we hoped to achieve and a set of means that we used to achieve these, some explicit and some implicit. The fact that one of the author's houses was struck by lightning three months before the end was certainly an unformed unknown.

At the start we anticipated getting further and establishing more. Its current point is complete, but the vision of what it could be has not been achieved. The task itself appears trivial and mechanistic, and seeing the difficulties is the first step in moving on. Why is it so difficult to understand clients? Why is it not possible to compare clients adequately? What we have found is that in order to start really understanding clients, we need much more background knowledge than might be expected. Every client as an organisation and as people is different; therefore, as we collect information, it is easy to see the differences between clients. Our attempt to generate a unified understanding is partly successful. We think that this is a tremendous achievement, because for the first time it lets the industry develop some generic client-handling skills. That this is not complete may not be surprising and indeed may not matter. It also allows others to challenge us to improve on this and show us better approaches to clients, if not least through criticising our attempts.

Our end goal was to be able to encapsulate each client sector into a chapter such that others could understand them in a way that would be useful to the construction industry. To do this, we needed to develop a model of enquiry, which in the end became our model of understanding.

Others simply describe features of clients (scale, expenditure, etc.). We sought a better way to describe clients so that we understand why they act in the ways that they do. We wanted to get into the minds of clients. The development of this model was through our experience rather than through a comprehensive exploration of the literature. This is not *the* model of clients but merely *a* model of clients that met our purposes.

Thus, the model focuses intentionally on certain areas of the client from our perspective. It is extremely complex, in that it requires a considerable amount of background knowledge in order to understand it. This makes it unwieldy for practical application and unapproachable by many within the industry. We apologise for this, as it reduces the model's usefulness. At this stage, it has not been possible to simplify the model, and we would certainly like this to happen.

Our means did involve a considerable amount of simplification of clients. It is possible for many to be critical of this. The problem is that in presenting sector information, it can appear naïve (in the sense that everyone knows what has been presented and the issues are not the most important aspect of the client), incomprehensible or incomplete. We recognise this as a problem and accept the criticism. There is, however, a problem of engaging all of our potential readers in our desire to address both an industry and an academic audience. Someone from the client organisation would have considerable knowledge, and all we can do is say that we are trying to provide a structure for this. For others, it is difficult to know the level at which to pitch. For some, it may be incomprehensible because we have not been able to simplify it sufficiently. Our approach has been to try to achieve a degree of comprehensiveness and connectedness. This drove us to seek a logical structure for the information, which of course elevates some issues above others. It also required us to take a particular view of how the information fitted together. This returns us to the model that was the basis of our enquiry and the structure through which we undertook the interviews and reported the results.

What we were doing then was model- or theory-building based on a phenomenological approach to research (Easterby-Smith *et al.*, 2002). This was undertaken in a social constructivist way in the sense that we belief society creates the situations that are found from the inside, rather than them being placed there from the outside. The situations are not independent of the observer or of society in general. The use of this model as the basis of the interviews delivers information that fits into the model. This is not a critical examination of the model and certainly not a testing of it. The detail may have errors, and there may be inadequate justification of many aspects. We believe that the value of the model as a whole gives it potential for powerful insight into the client–industry engagement. We hope that others will criticise it so that it can be justified better.

We are aware that the language we use for particular clients may appear naive to the clients themselves. Within a sector, certain words and phrases are a form of shorthand with particular meanings. The use of words also provides a shared association that allows a feeling of understanding and empathy. For example, in the supermarket sector, the word 'offer' means the range of goods that are offered for sale. To the layperson, the word 'offer' is more to do with a deal. It has not been possible to encapsulate this language. In the sector chapters, we have expanded the meaning of the words to make them understandable; this loses the specialised meaning of them and makes them look alien to people in the sector. We could have done this in a better way so that people could access a language set from a sector, which would allow them to align with the sector more rapidly. This is a future challenge.

In our research, we investigated other sectors, undertaking interviews, reading texts and acquiring company information, but we have reported on only six sectors as representative of clients. The other sectors informed our study, but including them would have made the book too long. We will seek other opportunities to disseminate these. We recognise that the sectors are not representative, as this is not possible. Our breakdown into private sector, public sector and mixed sector made the choice of two chapters in each of these classifications logical. There are many other sectors and subsectors within the clients, which we have not investigated. We are also aware that we have only investigated large clients. This is a serious omission as smaller clients create the greatest number of projects. The business constitution of smaller clients has much more variety, is less clear, and is more volatile than the large ones chosen. We hope that others may use the model to investigate these. Differences occur within sectors and even between individual clients in the same business. The industry has to learn to deal with this, and we believe that the model helps.

One feature we are acutely aware of is that these are all UK clients; indeed, most are English clients. Again, we believe that differences of country or region even are important. Our choice was pragmatic: we live in the West Midlands and investigated the easiest examples. We

	Name	Organisation
Schools	Chris Taylor	Bracknell District Council
	Peter Farrell	Birmingham City Council
	John Giacomelli	Staffordshire County Council
	Clare Collins	Staffordshire County Council
	Andrew George	Staffordshire County Council
	Ian Stafford	Wolverhampton Grammar School
Academia	Anne Hill	University of Central England, Birmingham
	John Kirk	University of Central England, Birmingham
	Matthew Smith	University of Central England, Birmingham

Author Index

Subject Index